T0313155

Handbook on ICT in Developing Countries: Next Generation ICT Technologies

RIVER PUBLISHERS SERIES IN COMMUNICATIONS

Series Editors:

ABBAS JAMALIPOUR
The University of Sydney
Australia

MARINA RUGGIERI
University of Rome Tor Vergata
Italy

JUNSHAN ZHANG
Arizona State University
USA

Indexing: All books published in this series are submitted to the Web of Science Book Citation Index (BkCI), to SCOPUS, to CrossRef and to Google Scholar for evaluation and indexing.

The "River Publishers Series in Communications" is a series of comprehensive academic and professional books which focus on communication and network systems. Topics range from the theory and use of systems involving all terminals, computers, and information processors to wired and wireless networks and network layouts, protocols, architectures, and implementations. Also covered are developments stemming from new market demands in systems, products, and technologies such as personal communications services, multimedia systems, enterprise networks, and optical communications.

The series includes research monographs, edited volumes, handbooks and textbooks, providing professionals, researchers, educators, and advanced students in the field with an invaluable insight into the latest research and developments.

For a list of other books in this series, visit www.riverpublishers.com

Handbook on ICT in Developing Countries: Next Generation ICT Technologies

Editors

Knud Erik Skouby

Idongesit Williams

Albert Gyamfi

Aalborg University
Denmark

LONDON AND NEW YORK

Published 2019 by River Publishers
River Publishers
Alsbjergvej 10, 9260 Gistrup, Denmark
www.riverpublishers.com

Distributed exclusively by Routledge
4 Park Square, Milton Park, Abingdon, Oxon OX14 4RN
605 Third Avenue, New York, NY 10158

First published in paperback 2024

Handbook on ICT in Developing Countries: Next Generation ICT Technologies / by Knud Erik Skouby, Idongesit Williams, Albert Gyamfi.

Routledge is an imprint of the Taylor & Francis Group, an informa business

Publisher's Note
The publisher has gone to great lengths to ensure the quality of this reprint but points out that some imperfections in the original copies may be apparent.

While every effort is made to provide dependable information, the publisher, authors, and editors cannot be held responsible for any errors or omissions.

ISBN: 978-87-7022-098-9 (hbk)
ISBN: 978-87-7004-359-5 (pbk)
ISBN: 978-1-003-34604-3 (ebk)

DOI: 10.1201/9781003346043

Contents

Foreword   xiii

Preface   xv

List of Contributors   xvii

List of Figures   xix

List of Tables   xxi

List of Abbreviations   xxiii

Prospects and Challenges of Next Generation ICT Technologies in Developing Countries   **1**

Knud Erik Skouby, Idongesit Williams and Albert Gyamfi
References 8

1 A Regulatory Taxonomy for Cloud Services in Latin America   **11**

Silvia Elaluf-Calderwood
1.1   Digital Agenda, Digitalization in the Latin American Context 11
1.2   Cloud Architecture and Protection of Users Data 15
     1.2.1   Cloud Numbers in Latin America 16
     1.2.2   Relevant Questions When Looking to National Regulatory Practices 18
     1.2.3   Argentina 19
     1.2.4   Brazil 21
     1.2.5   Chile 22
     1.2.6   Mexico 23

1.2.7   Colombia   24

1.2.8   Peru   25

1.3   Challenges to the Expansion of Cloud Services in Latin
America   26

1.3.1   Data Management   26

1.3.2   Expansion of access to services   27

1.3.3   Future of Data, Artificial Intelligence (AI), Internet
of Things (IoT) and the Creation of Employment in
Latin America   28

1.4   General Comments   30

References   32

2   **Lessons from Brazil's National Computer Policy for a New Industrial Policy on Industry 4.0**   **37**

Walter Shima, Pollyanna Rodrigues Gondin,
Marcelo Castellano Lopes and Marcelo Vargas

2.1   Introduction   38

2.2   The State as an Articulator and Policy Promoter   39

2.3   Industrial Policy and Technological Trajectory of Brazil's
Computer Industry   42

2.3.1   The Creation of the Computer Policy in the 1970s:
The Active Role of the State and the Institutional
Articulation   42

2.3.2   Computer Policy in the 1980s: The Attempt of
Intensifying the Industrial Policy and Strategic
Errors About the Technological Trajectory   45

2.3.3   The End of the Computer Policy from the 1990s:
Neoliberalism and the Dismantling of the State . . .   47

2.4   Perspectives of a Vertical Industrial Policy with Focus
on IOT   47

2.4.1   Evolution of CNAE Segments as Possible IOT
Developers   50

2.5   The Nature of the First Steps of Industrial Policy
of Industry 4.0   54

2.6   Conclusion   57

References   60

3 What Developing Countries Can Learn From The EU's GDPR 63

Roslyn Layton

3.1 Introduction 63
3.2 Theories of Data Protection 64
    3.2.1 Privacy vs. Data Protection 64
    3.2.2 Geopolitical Goals of the GDPR 65
    3.2.3 Data Protection and Cultural Norms 67
    3.2.4 Data Protection and Online Trust 68
3.3 Preliminary Outcomes 70
    3.3.1 Unintended Consequences 70
    3.3.2 Impacts to SMEs 73
    3.3.3 Compliance Costs 74
3.4 Conclusion 75
    References 76

4 Utility Cooperatives as Rural NGT Providers: Feasibility, Potentials and Pitfalls 85

Darío M. Goussal

4.1 Introduction 85
4.2 Background 87
4.3 The three Layers of Rural Broadband Feasibility 94
    4.3.1 Intrinsic Feasibility Analysis 95
    4.3.2 Institutional Feasibility 101
    4.3.3 The Third Layer: Willingness-to-Pay vs. Consumer Surplus 108
4.4 Conclusions 110
    References 113

5 Blockchain, Trust and Elections: A Proof of Concept for the Ghanaian National Elections 117

Idongesit Williams and Samuel Agbesi

5.1 Introduction 117
5.2 Electronic Voting, Technology and Trust 120
5.3 Blockchain and Election Process 123
    5.3.1 Overview of Blockchain Technology 123
    5.3.2 Potential for Blockchains and Elections 125
5.4 The Potential Use Case – The Ghanaian General Elections . 127
    5.4.1 The Ghanaian Election Process 127
    5.4.2 Challenges in the Ghanaian Election Process 129

5.5 Proof of Concept 130
    5.5.1 Current Trust Framework 130
    5.5.2 Potential Trust Framework with Blockchain 131
5.6 Conceptual Scenarios 132
    5.6.1 Scenario 1 (Remote Voting) 132
    5.6.2 Scenario 2 (Polling Booth Vote) 135
    5.6.3 Scenario 3 (Result Storage Only) 137
5.7 Discussion 137
5.8 Conclusion 140
    References 141

**6 Hybrid Cloud Adoption in a Developing Economy:
An Architectural Overview   147**
Kenneth Kwame Azumah
6.1 Introduction 148
6.2 Cloud Computing Background 149
    6.2.1 Service Models 150
    6.2.2 Deployment Models 151
    6.2.3 Considerations for Adopting Cloud Computing in an
          Enterprise 153
    6.2.4 Service Level Agreements 154
    6.2.5 State-of-the-Art in Hybrid Cloud Architecture . . . 154
    6.2.6 Requirements Engineering for the Hybrid Cloud . . 156
    6.2.7 Functional Requirements 156
    6.2.8 Non-functional Requirements 157
6.3 The Case of the Selected Hospital 158
    6.3.1 The Hospital's Information Systems 159
    6.3.2 Hospital Network Requirements and Business
          Constraints 159
    6.3.3 Functional Requirements Specification of the
          Hospital Network 160
    6.3.4 Non-functional Requirements of the Hospital
          Network 160
    6.3.5 Other Business Constraints 161
6.4 Infrastructure Redesign and Proposed Solutions 161
6.5 Measuring Performance Results 163
6.6 Results Discussion 164
6.7 Conclusion 167
    References 168

7   Developing Use Cases for Big Data Analytics: Data Integration with Social Media Metrics   **171**

Ezer Osei Yeboah-Boateng and Stephane Nwolley, Jnr

7.1   Introduction   172
    7.1.1   Problem Formulation   175
    7.1.2   Key Research Questions and Objectives   176
    7.1.3   Highlights of Findings   176
    7.1.4   Significance of the Chapter   177
    7.1.5   Outline of the Chapter   177
7.2   Literature Review   177
    7.2.1   Business Use Cases   178
    7.2.2   Data Integration Principles   179
    7.2.3   Big Data Analytics   180
    7.2.4   Social Media Metrics   185
7.3   Methodology   185
7.4   Social Media Metric Computations   186
    7.4.1   Volume   187
    7.4.2   Engagement   188
    7.4.3   Influence   189
    7.4.4   Reach   190
    7.4.5   Share of Voice   191
    7.4.6   Impact   191
7.5   Social Media Metrics and Use Cases   192
7.6   Social Media Metrics Computational and Data Integration Model   195
7.7   Conclusion   196
    References   198

8   Intrusion Detection and Prevention System for Wireless Sensor Network Using Machine Learning: A Comprehensive Survey and Discussion   **201**

Pankaj R. Chandre, Parikshit N. Mahalle, Geetanjali R. Shinde and Prashant S. Dhotre

8.1   Introduction   202
    8.1.1   Overview of the Problem   203
    8.1.2   How this Problem Affects India or Pune   204
    8.1.3   Why There is the Need for the Solutions You are Providing   204
    8.1.4   Why these Interventions/Intrusions are Important . .   205

8.2		Machine Learning	205
	8.2.1	Supervised Learning	205
	8.2.2	Unsupervised Learning	206
	8.2.3	Reinforcement Learning	206
	8.2.4	Semi-Supervised Learning	206
8.3		Motivation	206
8.4		Overview of IDS and IPS	207
	8.4.1	An IDS Overview	207
	8.4.2	IPS Overview	210
	8.4.3	Difference between IDS and IPS	212
	8.4.4	Threats	213
	8.4.5	How machine learning plays an essential role in IDS and IPS	216
8.5		Related Work and Gap Analysis	217
8.6		Attack Modelling	226
	8.6.1	Attacks at Network Layer	226
8.7		Issues and Challenges	230
	8.7.1	Why are Sensor Networks Difficult to Protect? . . .	230
	8.7.2	Issues in Wireless Sensor Network	230
	8.7.3	Challenges in Wireless Sensor Network	231
8.8		Proposed Work	231
8.9		A Need of Today	232
	8.9.1	What should be the Adverse Implications for not Adopting this Approach	232
	8.9.2	What should be done by Industry and Policy Makers?	233
8.10		Conclusions and Future Work	233
		References	234

9   Comprehensive Threat Analysis and Activity Modelling of Physical Layer Attacks in Internet of Things   237
Mahendra B. Salunke, Parikshit N. Mahalle and Prashant S. Dhotre
9.1		Introduction	238
	9.1.1	Overview	238
	9.1.2	IoT Based Business Use Cases	240
	9.1.3	Challenges in IoT	242
9.2		Motivation	243
9.3		Related Work	247
9.4		Gap Analysis	253

9.5   Threat Overview and Activity Modeling of Attacks   253
         9.5.1   Vulnerability vs. Threat vs. Attack   254
         9.5.2   IoT Reference Model   254
         9.5.3   Physical Layer Threats   255
9.6   Issues and Challenges   260
9.7   Proposed Methodology   262
9.8   Conclusion and Future Outlook   263
         References   264

10 An Extension of the Information Systems Success Model; A Study of District Health Information Management System (DHIMS II) in Ghana   269

Patrick Ohemeng Gyaase and Kodua Bright

10.1   Introduction   270
         10.1.1 Quality Characteristics of Health Information
                 System   271
         10.1.2 Challenges in Health Information Systems
                 Implementation   272
         10.1.3 The Conceptual Model for the Study; an Extended
                 Information Systems Success Model   272
10.2   Moderating Implementation Factors for the Success of
         Health Information Systems   274
10.3   Methodology   275
10.4   Data Analysis and Discussion of Results   275
         10.4.1 Demographic Data of Respondents   276
         10.4.2 An Analysis System Quality of DHIMS II   277
         10.4.3 An Analysis of Information Quality of DHIMS II .   277
         10.4.4 An Analysis of Service Quality of DHIMS II   278
         10.4.5 The Impact of Implementation Factors on the
                 Success of DHIMS II   279
10.5   Findings   280
         10.5.1 System Quality, Information Quality and Service
                 Quality of DHIMS II   281
         10.5.2 The Impact of the Extended Variable on DHIMS II
                 Success   282
10.6   Conclusions   282
         References   282

11 Reviewer Paper Assignment Problem – A Brief Review   **285**

Aboli H. Patil and Parikshit N. Mahalle

11.1 Introduction   285
11.2 An Overview of Existing Conference Management Systems   288
11.3 Related Work   289
11.4 Performance Evaluation Techniques and Data Sets Used For
      Testing   295
11.5 Brief Summary   296
      11.5.1 Research Opportunities – Gap and Challenges . . .   298
11.6 5G and Reviewer Assignment Problem   299
11.7 Conclusions   300
      References   300

Index   **303**

About the Authors   **309**

About the Editors   **317**

Foreword

The WWRF Series in Mobile Telecommunications

The Wireless World Research Forum (WWRF) is a global organization bringing together researchers from industry and academia to identify the key research challenges and opportunities across a wide range of aspects of communication technologies. Members and meeting participants work together to present their research and develop white papers and other publications to guide us towards the Wireless World. Much more information on the Forum, and details of its publication programme, are available on the WWRF website www.wwrf.ch. The scope of WWRF includes not just the study of novel radio technologies and the development of the core network, but also the way in which applications and services are developed, and the investigation of how to meet user needs and requirements.

WWRF's publication programme includes use of social media, online publication via our website and special issues of well-respected journals. In addition, where we have identified significant deserving subjects, WWRF is keen to support the publication of extended expositions of our material in book form, either singly authored or bringing together contributions from a number of authors. This series, published by River Publications, is focused on treating important concepts in some depth and bringing them to a wide readership in a timely way. Some will be based on extending existing white papers, while others are based on the output from WWRF-sponsored events or from proposals from individual members.

We believe that each volume of this series will be useful and informative to its readership, and will also contribute to further debate and contributions to WWRF and more widely.

Dr. Nigel Jefferies
WWRF Chairman

Professor Klaus David
WWRF Publications Chair

Preface

In recent years, there has increasingly been focus on ICTs in developing countries. ICTs are important enabling factors in several of UNs Sustainable Development Goals and universal accessibility to ICTs remains a serious concern. However, internet and mobile infrastructures in developing countries support an increasing stock of advanced information and communication applications in diverse fields of socio-economic activity including both the private and the public sector.

World Research Wireless Forum (WWRF) has focus on raising awareness and discussions on the potentials and barriers related to these applications based on wireless technologies. This book, the second in the series on ICTs in developing countries, is motivated by ongoing activities especially in Working Group A/B where the Next Generation Technologies are discussed from a user and business perspective. The issues are discussed both at a theoretical level and as cases in different countries across the continents.

The book addresses audiences within governments, organizations, industry and universities.

Sincerely thanks to authors: **Samuel Agbesi,** Center for Communication, Media and Information Technology (CMI), Aalborg University, Copenhagen; **Kenneth Azumah**, Center for Communication, Media and Information Technologies (CMI), Aalborg University, Copenhagen; **Ezer Osei Yeboah-Boateng,** Ghana Technology University College (GTUC); **Dr. Pankaj R. Chandre,** Department of Computer Engineering, STES's Smt. Kashibai Navale College of Engineering, Pune, India; **Dr. Prashant S. Dhotre,** Department of Computer Engineering, STES's Sinhgad Institute of Technology and Science, Pune, India; **Dr. Silvia Elaluf-Calderwood,** Florida International University USA & the University of Syracuse, NY, USA; **Dr. Pollyanna Rodrigues Gondin,** Federal University of Paraná, Brazil; **Prof. Darío M. Goussal,** Department of Electricity and Electronics, School of Engineering of Universidad Nacional del Nordeste (UNNE) at Resistencia, Argentina; **Dr. Patrick Ohemeng Gyaase,** Faculty of Information, Communication Sciences and Technology (ICST), Catholic

University College of Ghana, Fiapre, Sunyani; **Dr. Roslyn. Layton**, Center for Communication, Media, and Information Technologies (CMI), Aalborg University in Copenhagen, Denmark & the American Enterprise Institute; **Marcelo Castellano Lopes**, Public Policies at Federal University of Paraná, Brazil; **Dr. Parikshit N. Mahalle**, Department of Computer Engineering at STES's Smt. Kashibai Navale College of Engineering, Pune, India; **Dr. Stephane Nwolle,** Computer Science Department, Ashesi University; **Aboli H. Patil**, Smt. Kashibai Navale College of Engineering, Pune, India; **Mahendra B. Salunke,** Department of Computer Engineering, Pimpri Chinchwad College of Engineering and Research, Ravet, Pune; **Dr. Walter Shima,** Public Policies at Federal University of Paraná, Brazil; **Dr. Geetanjali R. Shinde,** Department of Computer Engineering, STES's Smt. Kashibai Navale College of Engineering, Pune, India; **Prof. Marcelo Vargas,** University of Paraná State, Brazil; **Dr. Idongesit Williams,** Center for Communication, Media, and Information Technologies (CMI), Aalborg University Copenhagen, Denmark.

The editorial team is grateful for inspiration and support from the publishers and WWRF.

I, finally, but not least thank Dr. Albert Gyamfi and Dr. Idongesit Williams for the tireless efforts of in collecting the contributions and editing them.

Professor Knud Erik Skouby

List of Contributors

Aboli H. Patil, *Department of Computer Engineering, Sinhagad College of Engineering, Pune, India; E-mail: aboleee.patil@gmail.com*

Albert Gyamfi, *Aalborg University, Copenhagen, Denmark; E-mail: alberto@cmi.aau.dk*

Darío M. Goussal, *School of Engineering, Northeastern University at Resistencia, Argentina; E-mail: dgoussal@yahoo.com*

Ezer Osei Yeboah-Boateng, *Ghana Technology University College (GTUC), Ghana and EZiTech, Ghana; E-mail: eyeboah-boateng@gtuc.edu.gh*

Geetanjali R. Shinde, *Department of Computer Engineering, Savitribai Phule Pune University, Pune, India; E-mail: gr83gita@gmail.com*

Idongesit Williams, *Aalborg University, Copenhagen, Denmark E-mail: idong@es.aau.dk*

Kenneth Kwame Azumah, *Centre for Communications, Media and Info. Technologies, Aalborg University, Copenhagen, Denmark; E-mail: kka@cmi.aau.dk*

Knud Erik Skouby, *Aalborg University, Copenhagen, Denmark; E-mail: Skouby@cmi.aau.dk*

Kodua Bright, *Holy Family Hospital, Nkawkaw, Ghana; E-mail: brightkoduah@gmail.com*

Mahendra B. Salunke, *SKNCOE Research Center, SPPU, Pune, India; E-mail: msalunke@gmail.com*

Marcelo Castellano Lopes, *Federal University of Paraná (UFPR), Brazil; E-mail: marcelolopescl@gmail.com*

Marcelo Vargas, *Federal University of Paraná (UFPR), Brazil; E-mail: marcelo.vargas@unespar.edu.br*

Pankaj R. Chandre, *Department of Computer Engineering, Savitribai Phule Pune University, Pune, India; E-mail: pankajchandre30@gmail.com*

Parikshit N. Mahalle, *Department of Computer Engineering, Savitribai Phule Pune University, Pune, India; E-mail: aalborg.pnm@gmail.com*

Patrick Ohemeng Gyaase, *Faculty of Information and Communication Science & Technology Catholic University College of Ghana, Fiapre, Sunyani, Ghana; E-mail: pkog@cug.edu.gh*

Pollyanna Rodrigues Gondin, *Federal University of Paraná (UFPR), Brazil; E-mail: pollygondin@gmail.com*

Prashant S. Dhotre, *Department of Computer Engineering, Savitribai Phule Pune University, Pune, India; E-mail: prashantsdhotre@gmail.com*

Roslyn Layton, *Center for Communication, Media and Information Technologies, Aalborg University, Denmark; E-mail: roslyn@layton.dk*

Samuel Agbesi, *Aalborg University, Copenhagen, Denmark; E-mail: sa@es.aau.dk*

Silvia Elaluf-Calderwood, *Florida International University, Florida, USA; E-mail: Selalufc@fiu.edu*

Stephane Nwolley, Jnr, *Npontu Technologies, Ghana; E-mail: snwolley@gmail.com*

Walter Shima, *Federal University of Paraná (UFPR), Brazil; E-mail: waltershima@ufpr.br*

List of Figures

Figure 1.1 OECD recommendations on digital government strategies. 30

Figure 2.1 From Industry 1.0 to Industry 4.0. 49

Figure 4.1 Estimated cost breakdown in a prototyping rural GPON network in Argentina. 100

Figure 4.2 Costs of all services of a typical utility cooperative providing urban broadband (Argentina-Group B). . 105

Figure 4.3 Costs of connectivity of a typical utility cooperative serving only urban areas (Argentina-Group B). . . . 105

Figure 4.4 Profit/Loss in rural and urban broadband vs. electricity or telephony (Group "A" utilities). 106

Figure 4.5 Profit/Loss in urban broadband vs. electricity distribution (Group "B" utilities). 107

Figure 5.1 End-to-end Blockchain solution. 133

Figure 5.2 Access to both Blockchains on the Electoral Commission's portal. 134

Figure 5.3 Blockchain architecture interfaced with DRE or I-voting. 135

Figure 6.1 Contributing technologies into cloud computing. . . 151

Figure 6.2 Representation of the hybrid cloud architecture. . . 155

Figure 6.3 Layer-isation of cloud computing modules. 157

Figure 6.4 Hybrid Cloud to enhance availability and ensure regulatory compliance. 162

Figure 6.5 Response times as measured with fiddler. 164

Figure 7.1 Big Data Processes. 182

Figure 7.2 Social Media Metrics Computational and Virtual Data Integration Model. 196

Figure 8.1 Flow of an IDS. 209

Figure 8.2 Shows the working of IPS. 211

Figure 8.3 Sybil attack. 227

Figure 8.4 Wormhole attack. 228

Figure 8.5   Blackhole attack. 229

Figure 8.6   Denial of Service attack. 229

Figure 8.7   Proposed IPS architecture to prevent attacks. 232

Figure 9.1   Components of IoT. 239

Figure 9.2   Taxonomy of IoT use cases. 241

Figure 9.3   Virtual shopping scenario. 245

Figure 9.4   Four layer IoT reference model. 255

Figure 9.5   Denial of service attack. 256

Figure 9.6   Message replay attack. 257

Figure 9.7   Tag cloning attack. 258

Figure 9.8   Spoofing attack. 259

Figure 9.9   Eavesdropping. 259

Figure 9.10   Jamming attack. 260

Figure 9.11   Side-channel attack. 261

Figure 9.12   Proposed Methodology. 262

Figure 9.13   Proposed Methodology. 263

Figure 10.1   An Adaptation of IS Success Model. 273

Figure 10.2   Distribution of Respondents. 276

Figure 11.1   Current status count of Conferences scheduled worldwide for span of Nov 2018 to Dec 2019. . . . 286

Figure 11.2   Current status of contribution of countries in Asia in scheduling Conferences (Nov 2018 to Dec 2019). . 287

List of Tables

Table 1.1   List of the main Latin American telecommunications
             regulators   13
Table 1.2   Data privacy questions   18
Table 1.3   Security   19
Table 1.4   Cybercrime   19
Table 1.5   Intellectual property rights   19
Table 1.6   Standards and international harmonization   20
Table 1.7   Promoting free trade   20
Table 1.8   IT readiness, broadband deployment   20
Table 2.1   Number of Establishments in Division 26 –
             Manufacture of Computer, Electronic and Optical
             Equipment – 2006–2016   51
Table 2.2   Number of Establishments in Classes 6201-5;
             6202-3; 6203-1 – Software Development –
             2006–2016   51
Table 2.3   Number of Active Links in Division 26 – Manufacture
             of Computer, Electronic and Optical Equipment –
             2006–2016   52
Table 2.4   Number of Active Links in Classes 6201-5; 6202-3;
             6203-1 – Software Development – 2006–2016 . . .   52
Table 2.5   Degree of Education in Division 26 – Manufacture
             of Computer Equipment, Electronic and Optical
             Products – 2006–2016   53
Table 2.6   Degree of Schooling in Classes 6201-5; 6202-3;
             6203-1 – Software Development – 2006–2016 . . .   54
Table 4.1   USA Electric Cooperatives providing rural broadband
             services   88
Table 4.2   Other USA Electric Cooperatives providing rural
             broadband services   98
Table 4.3   Sample of utility cooperatives in Argentina with data
             recorded from balance sheet series   103

Table 5.1   Potential nodes for the federated Blockchain 132
Table 5.2   Prerequisites for scenario 1 138
Table 8.1   Difference between IDS and IPS 214
Table 8.2   Wireless Sensor Network layer-wise threats and countermeasures 215
Table 8.3   Gap analysis IDS and IPS 225
Table 9.1   Gap analysis of solutions on physical layer attacks . 253
Table 10.1 System Quality of DHIMS II 277
Table 10.2 Information Quality of DHIMS II 278
Table 10.3 Service Quality of DHIMS II 279
Table 10.4 Coefficients of independent variables 280
Table 11.1 Typical Assignment Problems 288
Table 11.2 Performance Measures popularly in use 295
Table 11.3 Data Sets used for performance testing 296

List of Abbreviations

4G	Fourth Generation Mobile
5G	Fifth Generation mobile
ACM	AnT Clustering Algorithm
ACO	AnT Colony Optimization
ADPR	Argentine Data Protection Regulations
ADSS	All Dielectric Self-Supporting
AE	Active Ethernet
AES	Advanced Encryption Standard
AI	Artificial Intelligence
API	Application programming Interface
ARPU	Average Revenue per User
ATSP	Asymmetric Travelling Salesman Problem
AU	African Union
AWS	Amazon Web Service
BCT	Block Chain Technologies
BNDES	Brazil's National Bank for Economic and Social Development
BR	Biometric Register
BVD	Biometric Verification Devices
CAGR	Compound Annual Growth rate
CAPEX	Capital Expenditure
CAPRE	Coordinating Commission for Electronic Processing Activities
CDN	Content Distribution Network
CEPAL	(Comisión Económica para América Latina y el Caribe) Economic Commission for Latin America and the Caribbean
CJ	Cooperative Jamming
CoI	Conflict of Interests
CPU	CPU
CSP	Cloud Service Providers
CSIDPS	Collaborative Smart Intrusion Detection Prevention System
CSCW	Computer-Supported Cooperative Work
DEH	Dedicated Energy Harvesting

DFSS   Design for Six Sigma (DFSS)
DHIMS   District Health Information Management System
DMADV   Define - Measure - Analyze - Design - Verify
DNS   Domain Name System
DOS   Denial of Service
DDOS   Distributed Denial of Service
DP   Data Protection
DPA   Differential Power Anal ysis
DRE   Direct Recording Electronics
DSM   Digital Single Market
e-health   Electronic Health
EC   Electoral commission
ECC   Elliptic Curve Cryptography
EDR   Extended Differential Reach
EMR   Electronic Medical Records and Billing
ETL   Extract, Transform and Load
EVM   Electronic Voting Machine
EU   European Union
FAC   Fully Allocated Costs
FCC   Federal Communications Commission
FIPS   Federal Information Processing Standards
FTTH   Fiber to the Home
FTTC   Fibre to the Cabinet
GA   Genetic Algorithms
GDPR   General Data Protection Regulation
GPS   Global Positioning System
GSM   Global System for Mobile Communications
GSMA   Global System for Mobile Communications Association
GPON   Gigabit Passive Optical Networks
HFC   Hybrid Fiber-Coaxial
HIS   Health Information System
HRAIS   Human Resource and Accounting Information Systems
HTML   Hypertext Markup Language
HTTP   Hypertext Transfer Protocol
IaaS   Infrastructure as a Service
ICT   Information and Communications Technology
ID   Identity
IMF   International Monetary Fund
IS   Information Systems

IT	Information Technology
ITS	Intelligent Transportation System
IDS	Intrusion Detection Interference System
IETF	Internet Engineering Task Force
IAPP	International Association of Privacy Professionals
ICANN	Internet Corporation for Assigned Names and Numbers
IoT	Internet of Things
IPS	Intrusion Prevention System
ISP	Internet Service Provider
IT	Information Technology
JSON	JavaScript Object Notation
KPI	Key Performance Indicators
LAC	Latin America and the Caribbean
LAN	Local Area Network
LSI	Latent Semantic Indexing
MAC	Media Access Control
M2M	Machine-To-Machine
Mbs	megabits per second
MIMO	Multiple-Input-Multiple-Output
MTBF	Mean Time Between Failures
MTTR	Mean Time To Recovery
MVA	Minimum Viable Architecture
NAS	Network Attached Storage
NDPDP	National Directorate for Personal Data Protection
NGT	Next Generation Technologies
NIDPS	Network Based Intrusion Detection Prevention System
NIPS	Neural Information Processing Systems
NS	Network Simulator
OECD	Organisation for Economic Co-operation and Development
ODN	Optical Distribution Network
OEH	Opportunistic Energy Harvesting
OLT	Optical Line Terminal
ONU	Optical Network Unit
OPEX	Operational Expenditures
OSI	Open Systems Interconnection
PaaS	Platform as a Service
PCC	Program Committee Chair
PDPL	Personal Data Protection Law
PKI	Public Key Infrastructure

PLIS	Pharmacy and Laboratory Information Systems
QoE	Quality of Experience
QoS	Quality of Service
RAM	Random Access Memory
RAP	Reviewer Assignment Problem
RDBMS	Relational Database Management System
REST	Representational State Transfer
RFID	Radio-frequency identification
RIS	Radiology Information System
RMA	Reliability, Maintainability and Availability
ROI	Return on Investment
RTO	Response Time Objective
SaaS	Software as a Service
SEO	Search Engine Optimization
SG	Smart Grid
SLA	Service Level Agreement
SME	Small and Medium-Sized Enterprise
SOAP	Simple Object Access Protocol
SPA	Simple Power Analysis
SQL	Structured Query Language
SSD	Solid State Drive
TCO	Total Cost Of Ownership
TPC	Technical Program Committee
TS	Tabu Search
UML	Unified Modeling Language
UN	United Nations
UPS	Uninterruptible Power Supply
USD	United States Dollars
VM	Virtual Machines
VSN	Virtual personal Network
VSM	Vector Space Modeling
WAN	Wireless Area Network
Wi-Fi	Wireless Fidelity
WSN	Wireless Sensor Network
WWW	World Wide Web
XML	eXtensible Markup Language

Prospects and Challenges of Next Generation ICT Technologies in Developing Countries

Knud Erik Skouby, Idongesit Williams and Albert Gyamfi

Aalborg University, Copenhagen, Denmark
Skouby@cmi.aau.dk, idong@es.aau.dk, alberto@cmi.aau.dk

This second volume of the Handbook of ICT in developing countries is about the potentials of Next Generation Technologies. The first volume was discussing drives and barriers in the implementation and service delivery of the forth-coming 5G networks. This volume focus on the Information and Communication Technologies (ICTs), whose service delivery capacity will be greatly enhanced with either 5G networks or broadband Internet networks. The ICTs discussed are identified as Next Generation Technologies or NGTs. Examples of such technologies include Artificial Intelligence (AI), Machine Learning, Augmented Reality, Internet of Things (IoT), Autonomous Driving, Blockchain solutions, Cloud solutions etc.

These technologies possess inherent technical possibilities that deliver value to businesses and citizens. With IoT and AI, businesses and citizens can facilitate ubiquitous connectivity and semantic interactivity between humans and artifacts resulting in, e.g., smart production, smart enterprises and smart homes. With Block chain solutions, data access, storage and information can be decentralized, shared and delivered efficiently in a trusted environment opening for smooth and transparent production and delivery chains for industry and institutions including both goods and services – including also, e.g., election processes as discussed. With augmented reality, companies can create simulation of their products without having to engage in costly, actual pilot developments. Some of these potential are already being harnessed using 4th generation (4G) mobile networks. 4G networks have the capacity to support not only human-to-human or human-to-machine communications,

1

but also machine-to-machine communications even if especially in relation to this, it is less efficient than the upcoming 5G technology. Based on the 4G capacity there has been growth and potential growth shown in the adoption of certain NGTs. The rate of IoT adoption illustrates this as it has increased from 15.41 billion devices connected devices in 2015 to 20.35 billion devices connected devices in 2017 (Statista, 2019a). It is estimated that by 2025, 75.44 billion devices will be connected (ibid).

Less actual implementation and more promises are associated to, e.g., AI and Blockchains. Currently, 5% of global organizations have fully adopted AI (Statista, 2019b). But the prospects of more adoption possibilities exists as 18% and 23% of global organizations have either partially adopted or have a pilot project on AI respectively (ibid). The penetration of AI is expected to increase rapidly between 2017 and 2025 (Statista, 2019c). Blockchains are increasingly being deployed or experimenting with in different sectors of the economy (Statista, 2019d).

It is also important to note that companies gain added value by using more than one of these technologies for the delivery of their services. In some cases, the services are delivered in heterogeneous mesh networks. An example is the combination of AI and IoT in the delivery of manufacturing services. This is evident in industry 4.0 manufacturing setups. Although, the values of these technologies has been enabled by 4G, 5G and next generation mobile networks will enhance these values. This is because these next generation mobile networks will provide better Quality of Service (QoS) and Quality of Experience (QoE) possibilities than the current 4G networks. They will deliver also deliver higher capacity, higher data rates, reduced latency and mobility possibilities than 4G. Hence, there is greater prospects for the delivery and growth of NGTs globally.

Despite the convincing arguments for growth and benefits of NGTs, it is not obvious that this growth will also occur in developing countries. Currently the digital divide between developing countries and developed countries is widening as NGTs are being introduced into the market (GSMA, 2018) (Belitski & Liversage, 2019) (Muhammed et al., 2017). This is mostly due to the lack of infrastructure and low level of awareness by the businesses and citizens of the value made possible by NGTs for developing countries (Takeuchi, 2017) (Access Partnership, 2018). Main issues are high cost of access and high cost of deployment of NGTs which will result in the low demand for NGTs. These NGT problems are an extension of the adoption issues with respect to Broadband network technologies and the supported services in developing countries. For example, it is evident that the cost of subscription to mobile broadband as compared to fixed broadband in developing country is low and relatively affordable. However, the cost component impeding the adoption of mobile broadband services is the relatively high cost of smart phones in developing countries (GSMA, 2018). Hence, the cost of access to mobile broadband becomes higher resulting in low demand for broadband, which will result in a widening digital divide. One could also say that the absorptive capacity of developing countries for Broadband is low. That trend is likely to continue with NGTs. Apart from these market related challenges, there is general challenge to the take-off of NGTs, which may be even more challenging for developing countries. This threat is related to cybersecurity and privacy.

Cybersecurity threats include unethical hacking, Denial of Service (DOS), social engineering, threat to privacy, phishing etc. (Abdullahi, 2014) (Blackman, 1998). Some of these threats such as unethical hacking and privacy related threats will affect especially AI and IOT (Abomhara & Køien, 2015). This is because these services actually require the Internet, which is a vulnerable infrastructure, to function. The breach of such networks could have dire consequences, especially if the service involves the use of robots as a service assistance. Organizations and citizens in less well-established Internet environments might be more likely to shy away from in it. If these challenges, crop up once 5G is deployed, they have the potential to impede the growth in the adoption of NGTs in developing countries.

Therefore, to facilitate growth in the adoption of NGTs in developing countries, there is need for market interventions that facilitates both the demand and the supply side of the market. Such interventions will include policy and regulatory interventions and can be market based or direct public interventions as the situation may demand. Interventions aimed at facilitating cyber security will have to be promoted and implemented as means of facilitating trust for the NGTs. In addition, there is need for service inspiration or use-cases for developing countries. This will serve as an inspiration for a greater uptake of NGTs especially by SMEs and enterprises located in rural areas. Therefore, in this book, there is a great emphasis on policy interventions and inspiration for the adoption of NGTs, not just by organizations and citizens but by governments as well.

Although the adoption of NGTs in developing countries is low, there are positives that point to the fact that interventions and inspirations will facilitate the growth of NGTs. As mentioned earlier, 5G and next generation mobile networks will provide technical efficiency to the deployment and delivery of NGTs. There is no doubt that developing countries are gearing up for 5G and NGTs. There have been ongoing discussions in developing countries in Asia, Africa and South America on how to allocate 5G spectrum (GSA, 2019) (Alertify, 2019). Mobile network operators in some developing countries are also pushing for 5G for example, in Ghana (Authur, 2018) and in Chile, where the network operator Entel has upgraded its network infrastructure to enable the deployment of 5G (Daniels, 2017). To enable the Machine-to-Machine (M2M) and people-to-machine communications market, mobile network operators are supporting NGT start-ups in Africa (GSMA, 2018). These network operators see 5G as a technology that will provide competitive advantage in the market, hence they are devising innovative ways of achieve the first movers advantage. In addition to the discussions and the bid by mobile network operators to stimulate the market, very few developing country based entrepreneurs are adopting NGTs and most of them are either implementing AI, IOT or both. For example, there are few AI-based start-ups in Argentina, Brazil, Chile, Kenya, China and India to mention a few (Nanalyze, 2019) (Maseko, 2017). They operate in the agricultural sector, marketing, tourism etc. There are also some start-ups delivering block chain services in developing countries (Aguyi, 2018) (Keleman, 2018). Augmented reality has not gained much traction yet in developing countries. These initiatives, though few indicate that there is potential in developing countries to adopt and utilize NGTs. Hence, the interventions and the inspirations proposed in this book will serve as a push that will enable growth in the adoption and utilization of NGTs.

The book is made up of 11 chapters and four thematic areas, two of which were mentioned earlier. These are the policy related analysis and NGT use cases or inspirations. The additional thematic areas are cybersecurity threats and solutions; and information system solutions that can be deployed using NGTs.

The policy analysis are presented in Chapter 1 to Chapter 4.

- Chapter 1 provides an analysis of the regulatory classification or taxonomies for cloud based services in Latin America. Currently there are cloud based IoT services such as Google Cloud, Amazon Web Services (AWS) etc. The advantage of cloud computing is that it is a technology that enables enterprises to deploy IT infrastructure, software and services in a cost effective way. Cloud based services enable the enterprise to buy access to platforms providing these infrastructures, software and services to deliver their solutions. In order to facilitate the growth in the delivery of these emerging cloud solutions, developing countries have to develop digital agenda's that will create conducive market conditions and cater for technical and social challenges that may impede citizens' adoption of cloud based NGT services. In this chapter, Latin America is identified as a region where there is massive potential for the growth of emergent cloud services. Despite these growth patterns, the industry is still plagued by challenges with respect to the management of user data, security of user data, expansion of access to services and the lack of the requisite manpower. These challenges have the potential to impede the growth in the delivery of cloud-based services in Latin America. The chapter highlights the significance of these problems and the potential policy related solutions.

- Chapter 2 is a critique of the Brazilian NGT policies, which is summarily referred to as the Industry 4.0 policy. This critique comes in light of Brazilian manufacturing companies embracing industry 4.0. The industry adoption of industry 4.0 is supported by initiatives from software giants such as Microsoft and IBM who have launched AI skills initiative and opening research centers in Brazil. The critique identifies a policy implementation pattern, which dates back to the defunct Brazilian computer policy. Based on this critique, the chapter offers solutions that would correct the approach to the ICT policy in Brazil, which will in turn have a positive effect in the implementation of industry 4.0 related policies.

- In Chapter 3, the focus is on the challenges from the EU General Data Protection Regulation (GDPR) to the take-up of NGTs by SMEs in developing countries. Trade from individual developing countries to the EU is quite small. However, the small number of trading companies who provide online services of some form to the EU have to comply to the GDPR. The cost of complying to the GDPR comes at a huge cost. This cost could close down an SME in a developing country. This chapter argues that these costs are here to stay and will become even more pronounced with the delivery of NGTs because certain NGT services will require harvesting of big data. These data in most cases will be personal data of end users of the service. The chapter proposes a policy based solution, policy makers from developing countries could adopt to protect their SMEs from defaulting the GDPR, which could have bad consequences for the SME.

- The focus of Chapter 4 is on the potential of utility cooperatives to deliver NGTs. As mentioned earlier, the digital gap between developed and developing countries is widening. This in part is because of the slow or stagnant diffusion of new ICT technologies and services in rural areas of developing countries. Most rural areas in developing countries are either agrarian communities or areas sustained by limited commerce. But NGTs, especially AI and IoT provides the technological possibility for the facilitation of service-based networks in rural areas. Such networks could include the delivery of smart grids, sensor enabled animal husbandry and crop agriculture, elements of smart homes services, and other relevant use-cases. In most cases, this will require the replacement of existing systems in these rural areas. However, these deployments may not be possible in some rural areas if the service provider is huge and only interested in markets of a certain size. Based on experience, it is proven that well organized cooperatives can deploy and manage the services in a cost effective manner (see (Williams, 2018) (Goussal, 2017)). The chapter is a policy input providing insight on how organized cooperatives can be licensed as NGT providers to facilitate the diffusion of the NGTs to rural areas.

The NGT use case inspirations are presented in Chapters 5 to 7.

- Enterprises are currently testing the feasibility of using blockchains in some or all of their operations,but blockchains is of interest not only to private organizations but also to public organizations as well. There are certain public services that require high level of trust in the service delivery process. An example is the conducting of free and fair elections, which eludes most developing countries. Chapter 5 is designed to simulate how blockchains can serve in facilitating a free and fair election process. The simulation is in the form of a proof of concept on the different scenarios where blockchains can be used to facilitate a national election. Ghana is uses as the case for the proof of concept.
- Chapter 6 provides an insight into how cloud based services can be deployed in health care systems in developing countries. The chapter provides scenarios on the architectural structure and technical requirements for the deployment of cloud based services in a hospital scenario. It further provides an insight on how to measure the performance of the cloud based service in order to ascertain the QoS of the service provided.
- Chapter 7 has its focus on the management of big data. There has been a rise in the production of big data over time (Iqbal & Nawaz, 2019). The advancement in next generation telecom and Internet networks and the next generation technologies has led to the escalation in the production of big data because organizations are now inundated with big data which

are a mix of a variety of voluminous structured and unstructured data sets. This makes the data accessed difficult to process using traditional databases. Although there are technologies such as Hadoop that enables the processing of big data, it is fraught with data sorting and handling limitations (ParAccel, 2012). These limitations often result in the over-looking of data that would have been the source of an innovation for enterprises. These deficiencies opens up the possibility for identification of new ways to handle of big data. The deficiencies do not enable the enterprises to fully maximize big data to attain the level of competitive advantage they desire. Based on data integration techniques, the chapter proposes using the social media metrics computational and virtual data integration model. This model will help future enterprises generating big data via NGTs to intelligently sort the data in a manner that will reveal the innovation that elude big data technologies.

The potential cyber security threats and solutions are presented in Chapters 8 and 9 as these are essential for realizing the potentials in the adoption of NGTs. This idea is also supported in the first chapter of this book. This section of the book provides an insight into possible cyber security challenges that might be encountered in the delivery of NGT services. It further provides an insight into how these challenges can be dealt with. Chapter wise:

- In Chapter 8, the focus is on the detection of intrusion and how machine-learning techniques can be used to detect intrusion.
- In Chapter 9, the focus is on threats at the physical layer of IoT networks and how these threats can be addressed.

The potential information systems that can be deployed using NGTs are presented in Chapters 10 and 11. The essence of this section is to point to values ICTs present to formal and informal organizations. This value will obviously grow with the adoption of NGTs.

- In Chapter 10, the approach is to evaluate the successful implementation of an Information System (IS) in a formal organization. The evaluation is performed based on the IS success model of Delone & McLean. The aim is to determine if this success model is enough to determine the success or failure of an IS. The chapter indicates that the variables systems quality, information quality and service quality are no more enough to ascertain the success of an IS. It further reveals that data integrity, user resistance and poor management will lead to the failure in the implementation of an IS system. The analysis in this chapter is a mirror of the potential success criteria that could affect the outputs of NGT services such as AI based services and machine learning based services because low data integrity in the NGT services are bound to result in the rejection of the service being delivered.

- Chapter 11 provides a literature review of existing academic conference organization platforms. These platforms serve mostly informal organizations. Inadequacies are especially found in relation to management of reviewer assignments. These inadequacies affect the Quality of Service of the platforms. The chapter identifies platform, network and service delivery challenges in these platforms. The chapter calls for the use of NGTs such as AI, Block chain and emerging ICT ecosystems to fix the platform problems. It further calls for the adoption of 5G networks to aid in facilitating the Quality of Service lacking on most of the platforms.

The handbook is designed for a broad audience including practitioners, researchers, academics, policy makers and industry players and influencers. The language and approach to the handbook is a combination of the academic writing style and professional reviews. The idea behind this book is to provide information, inspiration, and advice on how to advance the growth in adoption of NGTs in developing countries.

References

Abdullahi, I. (2014). Mobile devices vulnerabilities: Challenges to mobile development in Africa. in K. E. Skouby and I. Williams (eds). The African Mobile Story, Aalborg, River Publishers, pp. 79–93.

Abomhara, M. and Køien, G. M. (2015). Cyber Security and the Internet of Things: Vulnerabilities, Threats, Intruders and Attacks. *Journal of Cyber Security and Mobility*, 4(1).

Access Partnership. Artificial Intelligence for Africa: An Opportunity for Growth, Development, and Democratisation. University of Pretoria, Pretoria, 2018.

Aguyi, W. (2018). 7 African Blockchain Startups to Watch in 2019. Available at: https://bitcoinafrica.io/2018/12/14/african-blockchain-startups/.

Alertify. (2019). 5G Spectrum In 2018 — Report By GSA.

Authur, A. (2018). Airtel Tigo joins calls for 5G network in Ghana. Available at: http://citifmonline.com/2018/03/29/airtel-tigo-joins-calls-5g-network-ghana/.

Belitski, M. and Liversage, B. (2019). E-Leadership in small and medium-sized enterprises in the developing world. *Technology Innovation Management Review,* 9(1), pp. 64–74.

Blackman, R. (1998). Convergence between telecommunications and other media: How should regulation adapt? *Telecommunications Policy*, 22(3), p. 163–170.

Daniels, G. (2017). Ericsson Upgrades Entel's network in Chile, as vendors position themselves for 5G contracts. Available at: https://www.telecomtv.com/content/5g/ericsson-upgrades-entel-s-network-in-chile-as-vendors-position-themselves-for-5g-contracts-16127/.

Goal, A. (2018). Increased Adoption of Augmented Reality In Enterprise. Available at: https://dzone.com/articles/increased-adoption-of-augmented-reality-using-an-l.

Goussal, D. M. (2017). Rural Broadband in Developing Regions: Alternative research agendas for the 5G Era. In K. E. Skouby, I. Williams and A. Gyamfi (eds) *Handbook on ICT in Developing Countries: 5G perspective*, Gistrup, River Publishers, pp. 235–275.

GSA. (2019). Spectrum for 5G Networks: Licensing Developments Worldwide. Global Mobile Suppliers Association, Farnham, 2019.

GSMA. (2018). The Mobile Economy 2018. Available at: https://www.gsma.com/mobileeconomy/wp-content/uploads/2018/02/The-Mobile-Economy-Global-2018.pdf.

Iqbal, Q. and Nawaz, R. (2019). Rife Information Pollution (infollution) and virtual organizations in industry 4.0: Within reality causes and consequences, in *Big data and knowledge shaing in virtual organizations*, Hershey PA, IGI Global, pp. 117–135.

Keleman, A. (2018). A glance at the state of Blockchain in Latin America. Available at: https://blog.neufund.org/a-glance-at-the-state-of-blockchain-in-latin-america-aac3ce148c04.

Maseko, F. (2017). Top 10 African Internet of Things startups to watch. Available at: https://www.itnewsafrica.com/2017/12/top-10-african-internet-of-things-startups-to-watch/.

Muhammed, S. H., Salahudeen, A. S., Aliyu, S. M., and Mustapha, A. (2017). The Need to Accelerate Cloud Adoption in Developing Countries of Africa. *Journal of Computer Science Engineering and Software Testing*, 3(1).

Nanalyze. (2019). Top-10 Artificial Intelligence Startups in South America. Available at: https://www.nanalyze.com/2019/01/artificial-intelligence-south-america/.

ParAccel. (2012). Hadoop's Limitations for Big Data Analytics, *ParAccel.*

Statista. (2019a). Internet of Things (IoT) connected devices installed base worldwide from 2015 to 2025 (in billions). Available at: https://www.sta tista.com/statistics/471264/iot-number-of-connected-devices-worldwide/.

Statista. (2019b). Adoption level of artificial intelligence (AI) in business organizations worldwide, as of 2017. Available at: https://www.sta tista.com/statistics/747790/worldwide-level-of-ai-adoption-business/.

Statista. (2019c). Growth of the artificial intelligence (AI) market worldwide from 2017 to 2025. Available at: https://www.statista.com/statistics/607960/worldwide-artificial-intelligence-market-growth/.

Stastista. (2019d). Blockchain adoption phases in organizations worldwide as of April 2018, by industry*. Available at: https://www.statista.com/statist ics/878748/worldwide-production-phase-blockchain-technology-industry/.

Takeuchi, T. (2017). *How Africa can gain benefits from Next Generation Networks.* in K. E. Skouby, I. Williams and A. Gyamfi. Handbook on ICT in Developing Countries: 5G perspective, Gistrup, River publishers, pp. 211–233.

Williams, I. (2018). Community Based networks and 5G Wifi. *Ekonomiczne Problemy Uslug,* vol. 131, pp. 321–334.

World Bank. (2016). World Development Report 2016: Digital. World Bank, Washington.

1

A Regulatory Taxonomy for Cloud Services in Latin America

Silvia Elaluf-Calderwood

Florida International University, Florida, USA
Selalufc@fiu.edu

This chapter is an overview of the state of regulatory practices in the Latin American context for the emergent cloud services in the region. The expansion and rapid evolution of digital services has had an impact on the development of digital services all around the world. Over the next 10 years, the establishment of 5G networks in the region will drive these policies towards wider acceptance and evolution. This particular chapter identifies the current regulators and the common grouping of the digital agendas established by the countries in the region, which aim to increase digital participation of the citizens. We also aim to reflect on the impact of the different regimes active for protection of user data. Our analysis focuses primary on three sectors: banking, health care and e-governance. We take the countries leading in digitalization: Argentina, Brazil, Chile, Mexico, and we identify some of the challenges from the expansion of cloud services in Latin America to countries such as Colombia and Peru. We conclude the chapter with general comments on the future of digitalization in the region and some recommendations to regulators to speed up the process.

1.1 Digital Agenda, Digitalization in the Latin American Context

The term "Digital Agenda" is used by a diverse myriad of individuals, institutions and organizations to assess evaluate and plan actions aiming for the creation of digital societies and digital economies. The term is in use worldwide and the United Nations Secretary General (UN Secretary General, 2018). In a document entitled "Strategy on New Technologies", the term links the expansion of digital agendas to the 2030 Sustainable Development Agenda, which aims to facilitate their alignment with the values enshrined in the UN Charter, the Universal Declaration of Human Rights and the norms and standards of International Laws.

In the case of Latin America, many of the big goals of digitalization are prospective and open to huge potential if developed as per the plans proposed. The region itself has many challenges. Traditionally the cycles of development in the region have been linked to the export of commodities, but low productivity rates have caused a huge digital divide with negative contribution of many traditional sectors to the national PBIs. Perhaps the most alarming aspect of this overall development points to the need to close the digital divide and to create the necessary investment to provide connectivity for access to digital services to international standards; these actions will enable an increase in productivity and the emergence of non-traditional sectors for job creation (Bello, 2018).

Reports from the OECD (OECD, 2016) and the GSM Association (GSMA, 2017) show a considerable increase in the number of users of the Internet, as well as – following world trends – a reduction in the fixed line implementations and a faster increase in the mobile market. There are many other development gaps threatening the future development of Latin America. CEPAL for example gives the number of households not using the Internet as up to 66% in the region compared to 55% in the Pacific Alliance (CEPAL, 2017). Other areas such as Internet of Things (IoT) show even bigger gaps, such as Cisco which by 2021 will have levels of adoption of IoT below 5% compared to over 10% in USA and nearly that value in the EU (CISCO Systems, 2016.

To add to this difficult situation the ARPU for mobile is one of the lowest in the world (9.6 USD/month, compared to 49.1 USD/month in USA or 31.8 USD/month in EU) and economic indicators indicate a slow increase for the next five years (Analysis Mason, 2017). So there is a low increase in income, a low return per user, a heavy taxation, scarce 4G–5G spectrum, and perhaps most importantly there is an intention to enable competition between operators with regulation that is asymmetrical but also increases operational costs.

In this difficult economic context, governments of the region do have ambitious plans for the digitalization of the region, which in the area of digital economy include the creation of a regional digital market, digitalization of production and a taxation framework for the digital economy. This is together with the enhanced presence of digital government through the opening of a best practice for the private sector, open government and mobile government.

Aiming at this digital connectivity, the generation of new public policies and regulations, plus a goal to update and harmonize regulation towards a convergence and solve any pending issues in the spectrum, most of the countries in the region aim for a digital ecosystem that provides digital trust and cybersecurity, education and training for human resources, and gender equality. Additionally these policies of access are being tested with the issues related to net neutrality and the provision of economically accessible services.

However when the author of this article researched how such goals were actually being supported in each country, she found a wide range of levels of development and understanding of what is required and when and how to achieve such goals.

Just to give an overview of the many institutions providing regulations for the region, see Table 1.1. This table is updated as of October 2018. Most of the institutions listed in the Table 1.1 are heterogeneous in size and mandate. Some are institutional legacies from colonial authorities while others have been formed and reformed over the years. The reforms depend upon political agendas of the government holding the political power in the country of origin.

Table 1.1 List of the main Latin American telecommunications regulators

Country	Regulator(s)	Website
Argentina	Ente Nacional de Comunicaciones (ENACOM)	https://www.enacom.gob.ar/
Bolivia	Autoridad de Regulacion and Fiscalizacion de Telecomunicaciones y Transporte (ATT)	https://www.att.gob.bo/
Brazil	Agência Nacional de Telecomunicações (ANATEL)	http://www.anatel.gov.br/institucional/
Chile	Subsecretaria de Telecommunicationes (SUBTEL)	https://www.subtel.gob.cl/
Colombia	Comision de Regulaciones de Colombia(CRC)	https://www.crcom.gov.co/es/pagina/inicio

(Continued)

Table 1.1   *(Continued)*

Country	Regulator(s)	Website
Costa Rica	Superintendencia de Comunicaciones (SUTEL)	https://sutel.go.cr/
Cuba	Ministerio de Telecommunicaciones	http://www.mincom.gob.cu/
Dominican Republic	Instituto Dominicano de las Telecomunicaciones (INDOTEL)	https://www.indotel.gob.do/
Ecuador	Agencia de Regulacion y Control de las Telecomunicaciones (ARCOTEL)	http://www.arcotel.gob.ec/
El Salvador	Superintendencia General de Electricidad y Telecomunicaciones (SIGET)	https://www.siget.gob.sv/
Guyana	Public Utilities Commision (PUC)	http://puc.org.gy/index.php
Guatemala	Superintendencia de Telecomunicaciones (SIT)	https://sit.gob.gt/
Haiti	Conseil National des Télécommunications (CONATEL)	http://conatel.gouv.ht/
Honduras	Comision Nacional de Telecomunicaciones	http://www.conatel.gob.hn/
Mexico	Instituto Federal de Telecomunicaciones	http://www.ift.org.mx/
Nicaragua	Instituto Nicaraguense de Telecomunications y Correos (telcor ente regulador)	http://www.telcor.gob.ni/Default.asp
Panama	Autoridad Nacional de los Servicios Publicos (ASEP)	https://www.asep.gob.pa/
Paraguay	Comision Nacional de Telecomunicaciones (CONATEL)	https://www.conatel.gov.py/
Peru	Organismo Supervisor de la Inversion Privada en Telecomunicaciones (OSIPTEL)	http://www.osiptel.gob.pe/
Puerto Rico	Junta Reglamentadora de Telecomunicaciones de Puerto Rico	http://www.jrtpr.gobierno.pr/
Sint Marteen	Bureau Telecommunications and Post Sint Marteen	https://www.sxmregulator.sx/
Uruguay	Unidad Reguladora de Servicios de Comunicaciones (URSEC)	http://www.ursec.gub.uy/
Venezuela	Comision Nacional de Telecomunicaciones (CONATEL)	http://www.conatel.gob.ve/

All these different entities with similar names are the ones that are working at different levels of instance on the many issues related to digitalization. There is also a heterogeneous history linked to the creation of the different regulators, with different mandates or priorities as the main goals of regulation.

Many are directly linked to the protection of the consumer, or explicitly to custodians of competitive markets, which aim to offer the best product to consumers at lower cost. For that reason, the leverage these institutions have to make big internet companies – providers of cloud services – comply with jurisdiction based norms is very limited since their leverage is based on reporting cases of abuse, fines and civil reparation to citizens.

Recent cases of breaches of data held by financial services firms (e.g. Chile (Kirk, 2018a), Argentina (BBC, 2017), Brazil (Paganini, 2018), Mexico (Kirk, 2018b) have opened up the public space to the discussion of data protection. Many of the countries affected by the data breaches have established some kind of breach notification system but this only applies to a few countries in the Latin American area. Lexology lists an overview of the ways each country is dealing with the data breaches (Lus Laboris, 2018). In the next section, we will discuss some of the current arguments on protection of users' data and give an overview of some of the countries' main positions.

1.2 Cloud Architecture and Protection of Users Data

As part of the digital agenda of many Latin American countries the adoption of cloud computing is growing and is seen as having the potential to diminish barriers to entry and to lower investment risks, including those associated with quick technological obsolescence. Diverse government agencies are providing Cloud computing enabled e-Government services and have become a driver for Internet usage in the population, mostly through cellular phones. Following a political agenda, Cloud computing, seen as a new technology model in the chain of innovations that have followed the evolution of the Internet, holds the potential of generating a significant impact in the economy and increasing welfare to citizens. The ambition to be perceived as a "modern and digital" country is providing the elements for a race to adopt the technology fast and with little reflection over its long-term impact on society.

The use of cloud implies the need for consistent and well-established regulations or norms for data management. It is interesting to mention that Latin American citizens do express a high level of concern over the use and management of their data. The most recent IPSOS survey shows the concern of 63% of users with regards to data ownership and data breaches (IPSOS, 2018). This concern however has not been mirrored in the political agenda of the region, neither in a clear determination to address the issues of data management in a convergent manner between all countries.

The result is a very diverse range of regulations if any at all. The commonality is a lack of protection and foresight on the impact of lacking data protection rules for data management. Paradoxically, Latin-American countries have started to worry about data protection matters, because of the importance of protecting the rights from the data subjects. Also companies see an interesting market to be developed: data storage, but in order to be a suitable market for that business, the country needs to have strong regulations on data protection. Although the banking sector is leading the review of the cybersecurity aspects of data management, other sectors such as health care records, or location based services data are hardly discussed at any level by politicians, regulators and the society in general.

1.2.1 Cloud Numbers in Latin America

Over 180,000 companies in Latin America use some form of public cloud infrastructure. Within this segment, 5% spend more than $20,000/month on cloud infrastructure, representing over 8,500 companies (Team Intricately, 2018). Grupo Globo is the top cloud spender in Latin America. This media giant in Brazil spends over $11 million/month on cloud services, $7 million of which goes toward AWS. The rest of their spending are distributed across Telefónica, América Móvil, and others. Mercado Livre, the second largest spender, is an Amazon-like ecommerce store. They deploy a multi-cloud strategy that leverages AWS, IBM Cloud, and Akamai.

Intricately, a marketing intelligence company for cloud service providers, estimates that $1 billion was spent in Q2 2018 on cloud infrastructure in Latin America, putting 2018 cloud spend at a $4+ billion annual run-rate. The provider with the largest market share, AWS, controls over 45% of the market and Microsoft Azure is a distant second with 18%. What is most impressive is Amazon's continued dominance in the face of growing competition from Microsoft, IBM, and Google. With all three providers announcing new data centre locations in Latin America, it can expect the distance between Amazon and its competitors to shrink as competitors continue to bridge the product maturity gap.

What may be even more impressive is Amazon's growth within Latin America's content delivery market. While Akamai continues to lead the market with 38% market share, Amazon's CloudFront product has made huge strides in gaining market share on Akamai and is currently rewarded with 28% of the market.

Akamai's 38% market share is distributed across just over 1,000 customers. In comparison, CloudFront has over six times the number of customers in Latin America but they spend much less. While 3% of Akamai's customers spend over $20,000/month, only 1% of CloudFront's customers do.

Despite these differences, each provider serves many of the same enterprises in Latin America. In addition, in some cases, such as with El País and Mercado Livre – two of the largest CDN buyers with multi-CDN strategies – and with Akamai and Amazon, they are neck and neck generating nearly $1 million/month from each customer.

Intricately's surveys of Latin America's content delivery market estimate a valuation of just over $600 million annually, representing 15% of the $4 billion cloud infrastructure market in Latin America.

A cloud infrastructure provider's potential for growth in a region is largely dependent on that region's level of connectivity – and in Latin America, connectivity is steadily rising.

With an estimated population of 630 million people, about 417 million people (66%) currently have access to the Internet. This is up from 384 million people in 2016 and accounts for 10.3% of Internet users in the world.

The region's largest telecommunication providers are Telefónica, América Móvil, and Oi. These companies are the leading providers in Latin America with the highest cloud market share among Latin American-based telecommunication providers.

In Latin America, a combination of legacy infrastructure, a regulatory environment that tends to favour the incumbents, and a historically low (but growing) demand for cloud services has enabled telecommunication providers to control a significant portion of the market.

An overview of the countries with significant cloud development in relation to data privacy laws. There are some major areas or sectors to look at when trying to understand the forest of regulations, decrees, laws, etc for each country.

1.2.2 Relevant Questions When Looking to National Regulatory Practices

While defining how to analyse the issues related to cloud, it is important to define some relevant sectors that define the status of cloud in each of the countries and the scope of what is or not protected. There are many ways to do this analysis; however, based on country Cloudscore cards we have identified seven sectors and a number of questions to ask in relation to data privacy, security, cybercrime, intellectual property rights, standards and international harmonization, the promotion of digital trade and broadband plans (Tables 1.2–1.8). Cybercrime is a hot topic in the sub-region with many government organizations actively engaging in policymaking aimed to reduce it. Therefore, each set of questions have different levels of relevance depending upon the local politics.

In many cases, we did not find answers to all these questions, but it is important to list the questions to have a fully comprehensive understanding of what is missing and what is already in place. In addition, as explained very few, if only two or three countries in the region can answer the listed questions with relevant information. Below we move on to describe the specifics for some of the countries that can answer many of the questions listed above.

Table 1.2   Data privacy questions

Pr1	Is a data protection law or regulation in place?
Pr2	What is the scope and coverage of the data protection law or regulation?
Pr3	Is a data protection authority in place?
Pr4	What is the nature of the data protection authority?
Pr5	Is the data protection authority enforcing the data protection law or regulation in an effective and transparent manner?
Pr6	Is the data protection law or regulation compatible with globally recognized frameworks that facilitate international data transfers?
Pr7	Are data controllers free from registration requirements?
Pr8	Are there cross-border data transfer requirements in place?
Pr9	Are cross-border data transfers free from arbitrary, unjustifiable, or disproportionate restrictions, such as national or sector-specific data or server localization requirements?
Pr10	Is there a personal data breach notification law or regulation?
Pr11	Are personal data breach notification requirements transparent, risk-based, and not overly prescriptive?
P12	Is an independent private right of action available for breaches of data privacy?

Table 1.3   Security

Sec1	Is there a national cybersecurity strategy in place?
Sec2	Is the national cybersecurity strategy current, comprehensive, and inclusive?
Sec3	Are there laws or appropriate guidance containing general security requirements for cloud service providers?
Sec4	Are laws or guidance on security requirements transparent, risk- based, and not overly prescriptive?
Sec5	Are there laws or appropriate guidance containing specific security audit requirements for cloud service providers that take account of international practice?
Sec6	Are international security standards, certification, and testing recognized as meeting local requirements?

Table 1.4   Cybercrime

Cyber1	Are cybercrime laws or regulations in place?
Cyber2	Are cybercrime laws or regulations consistent with the Budapest Convention on Cybercrime?
Cyber3	Do local laws and policies on law enforcement access to data avoid technology specific mandates or other barriers to the supply of security products and services?
Cyber4	Are arrangements in place for the cross-border exchange of data for law enforcement purposes that are transparent and fair?

Table 1.5   Intellectual property rights

Ip1	Are copyright laws or regulations in place that are consistent with international standards to protect cloud service providers?
Ipr2	Are copyright laws or regulations effectively enforced and implemented?
Ipr3	Is there clear legal protection against misappropriation of trade secrets?
Ipr4	Is the law or regulation on trade secrets effectively enforced?
Ipr5	Is there clear legal protection against the circumvention of Technological Protection Measures?
Ipr6	Are laws or regulations on the circumvention of Technological Protection Measures effectively enforced?
Ipr7	Are there clear legal protections in place for software-implemented inventions?
Ipr8	Are laws or regulations on the protection of software- implemented inventions effectively implemented?

1.2.3 Argentina

Argentina has been an early adopter of the European regulations (GDPR). The right to the protection of personal data is enshrined in Section 43 of the

Table 1.6   Standards and international harmonization

Sih1	Is there a regulatory body responsible for standards development for the country?
Sih2	Are international standards favoured over domestic standards?
Sih3	Does the government participate in international standard setting processes?
Sih4	Are e-commerce laws or regulations in place?
Sih5	What international instruments are the e-commerce laws or regulations based on?
Sih6	Is there a law or regulation that gives electronic signatures clear legal weight?
Sih7	Are cloud service providers free from mandatory interference or censoring?

Table 1.7   Promoting free trade

Pt1	Is a national strategy or platform in place to promote the development of cloud services and products?
Pt2	Are there any laws or policies in place that implement technology neutrality in government?
Pt3	Are cloud computing services able to operate free from laws or policies that either mandate or give preference to the use of certain products, services, standards, or technologies?
Pt4	Are cloud computing services able to operate free from laws, procurement policies, or licensing rules that discriminate based on the nationality of the vendor, developer, or service provider?
Pt5	Has the country signed and implemented international agreements that ensure the procurement of cloud services is free from discrimination?
Pt6	Are services delivered by cloud providers free from tariffs and other trade barriers?
Pt7	Are cloud computing services able to operate free from laws or policies that impose data localization requirements?

Table 1.8   IT readiness, broadband deployment

ITBr1	Is there a National Broadband Plan?
ITBr2	Is the National Broadband Plan being effectively implemented?
ITBr3	Are there laws or policies that regulate "net neutrality"?

Argentine National Constitution (Constitution). The current legal framework of the Argentine Data Protection Regulations (ADPR) is made up of the Constitution and:

- Personal Data Protection Law No. 25,326 (PDPL). Regulatory Decree of the PDPL No. 1558/2001 (DP Decree).

- "Do Not Call" National Registry Law No. 26,951; on the possibility of the data owner to opt-out of marketing and/or advertising campaigns carried out by telephone.
- Provisions issued by the National Directorate for Personal Data Protection (NDPDP) (for example, Provision NDPDP 4/2009). This authority oversees the privacy laws.

Over the years, the NDPDP has issued several provisions on different subjects. It has issued provisions ranging from the withdrawal or blockage right of the data owner that must be clearly publicised in any advertising communication (Provision No. 4/2009) to the way by which personal data can be collected by drones (Provision No. 20/2015).

All public and private sector databases must be registered with the National Commission for the Protection of Personal Data. The registration for all users is available online and fees apply. In addition, Regulation No. 60-E/2016 on international transfers of personal data provides model contract language for international data transfers to countries that do not provide adequate levels of protection. The Regulation also requires notification to the DNPDP when contracts are used other than those provided in the Regulation. This acts as a de facto registration requirement for some international data transfers (BSA, 2018). The proposed draft bill (issued in 2017) to amend the data protection law generally removes registration requirements, but the progress of this bill is uncertain.

1.2.4 Brazil

Although the use of personal data is regulated in most Western countries, there is a lack of regulation in this field in Brazil as privacy and data protection are treated as distinct concepts, although they are both derive from the right to privacy, which is enshrined as a constitutional principle. While privacy is regulated in the Brazilian Civil Code (Article 21), 2 data protection demands specific rules. In this regard, Law No. 12.965 (Brazil's Civil Rights Framework for the Internet) regulates aspects of data privacy within the framework of the Internet; however, it only applies to the collection, storage and use of data in the context of the Internet. Outside the context of the Internet, there are no statutes regulating data protection in general, although some sector-specific laws regulate the protection of personal data and there are also bills currently under consideration in Congress that aim to regulate the protection and treatment of personal data in general in the Brazilian territory.

In 2018 there was a major change in the Brazilian normative application of data protection laws after a consultation process with industry and other stakeholders, the result was that some of the restrictions for jurisdictional holding of digital data were removed for financial services (AWS, 2018).

1.2.5 Chile

The legal framework for data protection can be found in article 19 No. 4 of the Political Constitution of the Republic of Chile (Magliona et al., 2018), which guarantees the respect and protection of privacy and honour of the person and his or her family at a constitutional level. In addition, Chile has a dedicated data protection law, Law No. 19,628 on Privacy Protection, which was published in the Official Gazette on 28 August 1999 (the Law). The current Law is not based on any international instrument on privacy or data protection in force (such as the OECD guidelines, Directive 95/46/EC, EU General Data Protection Regulation or the European Convention on Human Rights and Fundamental Freedoms).

There is no special data protection authority in Chile; data protection oversight is addressed by general courts with general powers. A summary procedure is established by law if the person responsible for the personal data registry or bank fails to respond to a request for access, modification, elimination or blocking of personal data within two business days, or refuses a request on grounds other than the security of the nation or the national interest.

There is no data protection authority in Chile. A bill has been discussed in the Congress that will reform the whole data protection environment in the country and will create the first data protection authority in Chile.

Breaches of data protection caused by improper processing of data may eventually lead to fines determined by the Law (ranging from US$75 to US$760, or from US$760 to US$3,800 if the breach comes from financial data). Fines are viewed and determined in a summary procedure.

The Law establishes a general rule under which both non-monetary and monetary damages that result from wilful misconduct or negligence in the processing of personal data shall be compensated. In those cases, the civil judge, considering the circumstances of the case and the relevance of the facts, shall establish the amount of compensation reasonably.

The Data Protection Law does not cover interception of communications or monitoring and surveillance of individuals. Both matters are regulated by:

- Law No. 19,223 (the Computer Crime Law);
- Article 161-A, 369-ter, 411-octies of the Penal Code; and
- Articles 222 to 226 of the Criminal Code of Procedure.

The Data Protection Law does cover electronic marketing, in the sense of establishing that no authorisation is required to make electronic marketing when the information comes from sources available to the public (registries or collection of personal data, public or private, with unrestricted or unreserved access to the requesters).

1.2.6 Mexico

Mexico due to its proximity to the USA has been an early developer of data protection and privacy rules (Perezalonso Eguia, 2017). In 2013, Mexico created the National Institute of Transparency, Access to Information and Personal Data Protection (Instituto Nacional de Transparencia, Acceso a la Información y Protección de Datos Personales) (the "INAI").

Federal Law for the Protection of Personal Data in the Possession of Private Parties (the "LFPDPPP") supplemented by the Rules of the Federal Law for the Protection of Personal Data in the Possession of Private Parties (the "Regulation").

These laws were supplemented in January 2013 by the parameters for the proper improvement of the mandatory self-regulation schemes referred to in Article 44 of the Federal Law for the Protection of Personal Data in the Possession of Private Parties (the "Parameters for Mandatory Self-Regulation"), which are designed to establish rules, standards and procedures for the improvement and implementation of mandatory self-regulation for the protection of personal data.

The Regulation applies to processing:

- by an entity with an establishment in Mexico;
- outside of Mexico if conducted for controller in Mexico;
- where the Regulation is applicable by principles of international law; or
- Where the controller is based outside of Mexico but uses equipment in Mexico (other than for the purposes of transit).

Data protection legislation is applicable to private parties (both individuals and legal entities), except for credit information companies and individuals who collect and store information for personal or domestic use for a non-commercial purpose or without the intent to disclose such information.

"Personal data" is defined as any information relating to an identified or identifiable individual. However, the Regulation does not apply to information regarding: (i) legal entities; (ii) individuals acting as merchants or professionals; or (iii) basic work related contact details.

Consent may be express or implied. Express consent may be given verbally, in writing, through an electronic medium, or by unequivocal signs. Implied consent results from non-objection to a privacy notice provided to the data subject.

In certain circumstances, express consent must be obtained, for example the processing of financial information or sensitive personal data.

1.2.7 Colombia

Colombia has been defining a clear digital strategy that includes some comprehensive privacy and data protection regulations, linked to its path to be a fully OECD member (Meltzer and Perez Marulanda, 2016).

On October 18, 2012 the Government issued Law 1581 of 2012 (hereinafter referred as the "Data Protection Regulation") in which the general provisions for personal data protection were established, in order to protect principally the data subject and guarantee its rights. It is worth mentioning that the discussion in order to issue the law was preceded by a judgment from the Constitutional Court regarding the constitutionality of the law, and many of the matters regulated by it were interpreted by the constitutional judges (Sentence C-748 of 2011). Some of those matters are the handling of children's data, the applicability of the law to a legal person, among others. The Constitutional Court established that children and adolescents' personal data can be treated by those in charge or responsible for databases, as long as it does not jeopardize the prevalence of rights and unequivocally pursuits a superior interest.

The Data Protection Regulation establishes that its dispositions will be of no application to the personal data contained in the following databases:

- Personal or domestic,
- Security and national defence, and the prevention, detection, monitoring, and control of money laundering and terrorism financing,
- Intelligence and counter-intelligence,
- Journalistic information,
- Financial and credit information (Law 1266 of 2008), and
- Population censuses (Law 79 of 1993).

As we mentioned before, there is also a regulation regarding data protection for the financial sector (Law 1266 of 2008), the Data Protection Regulation, being the general disposition on data protection, establishing that its general principles will in any case apply to the financial and credit information databases.

Law 1581 includes the rights of data subjects and the duties from data controllers and data processors. Data subjects are able to exercise their rights at any time, even those related to their elimination from the database, regardless of the authorization that she/he gave for handling the personal data. Regarding sensitive data (such as data related with gender, sexual orientation, political and religious views, clinical history, among others), it is established that such data can only be handled in the cases contemplated by the law, which include, among others, that the data subject authorizes its use. The authorization given in this regard must be express and previous and it is important that it establishes the data to which the authorization applies. A great variety of aspects regarding data protection were left to be regulated by specific rules on which we will comment.

1.2.8 Peru

Peru has adopted the following legal framework for the protection of personal data which regulate the collection and processing of personal data (Panez et al., 2018):

- Law No. 29733, Personal Data Protection Law, issued on 21 June 2011 and published in the Official Gazette on 3 July 2011; Supreme Decree No. 003-2013-JUS, Regulations on Personal Data Protection Law, published in the Official Gazette on 22 March 2013 and fully enforceable from May 8th, 2015; and
- Directorial Resolution No. 019-2013-JUS/DGPDP, Guidelines on Security of Information, published on 22 March 2013.

These documents in the Spanish version can be found at the Peruvian Minister of Justice and Human Rights repository[1]. The Personal Data Protection Law and its regulations apply to any person or legal entity, public or private, processing personal data in Peruvian territory.

It is interesting to note that besides the countries mentioned, the rest of the countries in Latin America have a diverse and heterogeneous mix of laws and policies. Much of what has been legally prescribed is very normative and

---

[1]Relevant Peruvian legislation can be searched at: www.minjus.gob.pe/legislacion/

in the next section, we will discuss further the implications of such policy in hampering or benefiting the process of digitalization.

1.3 Challenges to the Expansion of Cloud Services in Latin America

The current race between Latin American countries to attract the data centre locations for the top Internet companies (Amazon, Google, Facebook, Microsoft) is already in place and very heated. However, governments and the politicians leading these races must be aware of the drawback of providing so many incentives to these companies and not a comprehensive reciprocity plan that benefits the country and their citizens.

1.3.1 Data Management

The questions provided in Section 1.2.2 are relevant when we examine other sectors besides the banking sector when it comes to data management in the cloud. For this chapter we have identified another four areas where the debate on how data management and privacy is not being regulated and remains open to market laws and abuse. These are:

- **Health:** In the absence of a comprehensive definition of personal data[2] used by all countries of the region at the same time, a major area that can benefit from the reduction in the cost of using cloud services and the streamlining of data processing is the health care sector (E-Health Reporter Latin America, 2018). Many hospitals in the public sector have plans to upload services and data to clouds for the first time; they lag behind the private sector, which in many instances has pioneered the use of the cloud. Most of the health providers see SaaS (Software as a service) as a predominant choice for cloud. Between professionals and management of health care, there is a low level of awareness on the vulnerability of data, which explains the zero political value given to the management of this type of data.
- **Bio Resources and Agriculture:** Latin American is a region of the world favoured with a variety of natural resources and labour intensive agricultural sectors. Many of the countries of the region are leading

---

[2]Latin America does not have a standarised definition of personal data. Compared this to the GDPR definition that includes attributes such as genetic, mental, cultural, economic or social information (Computer Weekly 2017).

exporters of agricultural products. However, productivity is low and automation and the use of technology for the control of the water resources and land management cycles could significantly improve the rate of productivity. Currently production of exports is very labour intensive and the use of technology at all stages of production can only improve such productivity indicators.

- **Manufacturing and mining:** With the exception of the four country leaders (Argentina, Brazil, Chile and Mexico) the situation of the manufacturing sector is very similar to the agriculture sector. The exception is mining where the need to use significant resources for the installation of extraction plants requires the adoption of new technology. Manufacturing, however, struggles to compete with cheaper and more cost-efficient products manufactured in China, Vietnam, etc. This is particularly true for the area of white goods and machinery. Many of these imported devices are already able to collect significant data sets and there are no regulations or rules in place to ascertain where this data is being held, who is accessing it and what it is being used for.

- **Infrastructure management and control:** One of the goals of digitalization is the automation of infrastructure control such as traffic signals in the public transport network, control of energy stations (e.g. thermal power centrals, hydraulic projects such as dams, or water network canals), water processing centres, ports security etc. It is true that many of the current services are partially provided by either the public or private sector with contracts that are comprehensive in terms of the physical and operational aspects of the activity. However, the collection of data on performance, security breaches, etc. is not covered by such contracts.

1.3.2 Expansion of access to services

In the abstract, we mentioned the huge opportunities that can arise from the quick adoption of 5G and broadband technologies. The region however is hampered by the disparity in access to broadband services. Investors would first need to roll out a comprehensive plan for the development of state of the art networks, requiring significant government investment.

Additionally there is limited and inadequate technical expertise to develop and manage cloud-computing solutions. It has been highlighted that in the Latin American region, since most of the population is young, programs for the development of high skilled training are not very accessible; neither is there a culture of business development or access to credit (OECD/CEPAL/CAF, 2017).

Access to technology – mobile and cloud – presents an opportunity to provide this population access to resources in business training and access to financial services aid to development business. In many cases, simplification of the processes for access and creation of employment or companies will help to maximise the opportunities the new use of technology brings in and the same can be said in relation to taxation and the public sector (Seco and Muñoz, 2018).

1.3.3 Future of Data, Artificial Intelligence (AI), Internet of Things (IoT) and the Creation of Employment in Latin America

Cloud computing is seen as a "fast-forward" way to create employment and drive digitalization. However the start up benefits of bringing data centres to a country are offset long term by the actual reduction in the number of people needed to maintain, upgrade and run such centres. Economic studies based on sector assumptions for London and San Francisco have shown that the number of jobs created directly linked to the activities of the data centres is very low (Liebenau et al., 2012). In Latin American, the huge incentives offered to cloud companies to bring data centres to the region are still to be compared to the job creation that is, or will be, created in the region through direct or indirect activities related to the data centres.

- **Big data considerations:** As seen in Section 1.2.2 the many questions that are aimed to evaluate the regulation or practices applied to the digital sector ought to provide a frame for the protection and nurturing of digital industries associated with big data. It is important that regulation is neither seen as hampering the development of these sectors nor as lax and unable to provide any type of control, mediation or enforcement between stakeholders.

  It is very important that regulators encourage transparency, as well as enabling and protecting cross-border data flows. Rules restricting the transfer of data and information across borders, however, do not accommodate the current realities of broadband-enabled computing. Big data processing facilities through clouds allow the usage of the data at incredible speeds in all sorts of applications and usages.

- **Artificial intelligence (AI) and Internet of Things (IoT):** Robots and technology are becoming more and more accessible. The quality-adjusted price of industrial robots decreased by 80% between 1990 and 2005. Automation of production processes will increase in the coming years. This will have a major impact on employment. McKinsey estimated that around 51% of jobs in the United States are susceptible to being replaced by automation by 2050 (McKinsey & Company, 2018). The World Bank estimates that this figure is at or around 57% in OECD economies (World Bank, 2016). At the global level, McKinsey estimates that automation could impact 1.1 billion employees and US$ 15.8 trillion a year in wages (McKinsey & Company, 2018).

    Latin America needs more and better skills. The education level required in technological and automated jobs are higher than in jobs that are being replaced by technology (McKinsey and Company, 2018 and Hamel et al., 2014). However, it is estimated that 13% of activities carried out by university-educated professionals could eventually be automated (McKinsey & Company, 2018).

    Although inequality has decreased in the Latin America and the Caribbean (LAC) area, there is concern that technology will undo this trend. The open market of innovation raised from IoT is already available to Latin American consumers through many appliances and devices that can be purchased freely. Smart televisions, kitchen devices, etc. are IoT ready with the aid of a simple Wi-Fi connection and provide smart information to their manufacturers. Lobbying and data issues from Internet companies that are international and the manufacturing sector of digital artefacts (China and South Korea) are not discussed in any political agenda within the region.

    The jobs of the future will be more flexible. On the one hand, employees will have more freedom to choose where and when they want to work. On the other hand, the new jobs will be more informal and temporary. This, together with the lack of regulations and audit capacity by government agencies will cause some uncertainty with respect to employment protection for these future workers. These structural changes will determine the skills needed to join the labour markets of the future. In fact, experts predict a considerable increase in demand in information systems and technology skills – e.g. programming – and soft skills, e.g. effective communication and customer service (Inter-American Development Bank (IDB), 2018).

1.4 General Comments

The development of cloud computing has some constraints: limited access to and low quality of broadband Internet, legal and regulatory frameworks that are weak or still under development, and additionally Service Level Agreements (SLAs) that are incomplete because of the shortcoming in regulation and the limited assessment of privacy and security (ECLAC, 2014).

For the region the OECD has proposed a triptych set of recommendations for the development of digital government strategies (OECD, 2015). See Figure 1.1. Our analysis of the six areas that are required to have a consistent cloud service regulation as explained in Section 1.2.2 is the tool to achieve the recommendations of the table.

There is a symbiotic relationship between the enabling and compliance of the many aspects of data privacy, security, cybercrime, intellectual property rights, standards and international harmonization whilst enabling free trade. The OECD recommendations are slow to be implemented because the main issue with these twelve points is the lack of leadership and vision in the region.

Our taxonomy and early research shows that it is difficult to find the methods to ensure a coherent and holistic approach to these policies over the whole geographical region. There is a lack of long-term vision for the role of the state in promoting both broadband development and Cloud computing adoption. The private sectors, providing sub-contracted cloud services to the

Openness and Engagement	Governance and Coordination	Capacities to Support Implementation
1. Openness, transparency and inclusiveness	5. Leadership and political commitment	9. Development of clear business cases
2. Engagement and participation in a multi-actor context in policy making and service delivery	6. Coherent use of digital technology across policy areas	10. Reinforced institutional capacities
3. Creation of a data driven culture	7. Effective organisational and governance frameworks to coordinate	11. Procurement of digital technologies
4. Protecting privacy and ensuring security		12. Legal and regulatory frameworks
	8. Strengthen international cooperation with other governments	
Creating Value Through the Use of ICT		
Non OECD members: Colombia, Costa Rica, Egypt, Kazakhstan, Lithuania, Morocco, Peru, Romania, Russia		

Figure 1.1   OECD recommendations on digital government strategies.[3]

---

[3]Source OECD elaboration based on the recommendation of the Council on Digital Government Strategies (2014).

government, are in a unique position to provide input to this long-term view of cloud adoption.

Latin America as a whole requires investment in digital education. In the last thirty years the region has seen a reduction in the efforts for increasing the educational level of the population. The use and access of technology for itself is not a replacement for educational programs that are fully comprehensive. All the countries in the region have the ambition to have a digital economy of some kind, but based on the experiences of the past, the digital presence requires a clear mandate on what it is going to achieve. There is a deficit on content creation vs. technological innovation in the region that is getting wider and can destroy the last 15 years of economic growth in the region that have reduced inequality and poverty.

The region is being recognized by the USA International Trade Commission as an area with market opportunities and in particular Brazil is identified as a very important digital trade partner (US International Trade Commission, 2014). It is important for the region to wake up to the challenge of digitalization in the 21st century. The adoption of cloud computing in Latin America is at an early stage, and decisions taken today by policy-makers and other stakeholders will influence the degree to which the citizens of particular nations and in the region as a whole will benefit from this technology in the short to medium term (Gutierrez and Korn, 2014). Twenty-first century data protection rules and policies, and whether they are designed with the flexibility to accommodate this transforming technology, will play an important role in facilitating cloud computing adoption and the benefits it produces for national competitiveness, i.e. the economic growth and long-term improvement of a society's standard of living resulting from improvements in national productivity and efficiency.

Hence the need for the region stakeholders in cloud computing to advance in the understanding of the management of related risks, promote service transparency, ensure data portability, facility interoperability between cloud services provides and services, provide adequate networks and capacity, clarify and improve accountability of all stakeholders and accelerate the modernization and harmonization of regulatory frameworks.

Acknowledgements

The author wishes to thank you the Institut Barcelona d'Estudis International (IBEI) and Telefónica for the award of the Cátedra Telefónica IBEI's Research Fellowship for 2018 that allowed me to follow up this research avenue. I also would like to thank you the Florida International University by opening up access to cloud service providers in the Latin American region.

References

Analysis Mason. (2017). Latin America Telecoms Market: Interim Forecast Update 2016–2021. London, UK: Analysis Mason. Available at: https://www.giiresearch.com/report/an317855-latin-america-telecoms-market-interim-forecast.html

AWS. (2018). AWS User Guide to Financial Services Regulations in Brazil – Brazilian National Monetary Council, Resolution 4,658 Seattle, USA: AWS, 2018. Available at: https://d1.awsstatic.com/whitepapers/compli ance/AWS_User_Guide_for_Financial_Services_in_Brazil.pdf

BBC. (2017). Equifax Had 'admin' as Login and Password in Argentina. BBC.com. BBC Technology (blog), Available at: https://www.bbc. com/news/technology-41257576

Bello, P. (2018). Desafio de la digitalización en la Alianza del Pacífico. *presented at the IPAE, Lima, Peru*, Available at: http://asiet.lat/?wp dmdl=6042

BSA. (2018). Country Cloud Computing Score Card: Argentina. BSA The Software Alliance, https://cloudscorecard.bsa.org/2018/pdf/ country_repo rts/2018_Country_Report_Argentina.pdf

CEPAL. (2017). Economics Survey of Latin American and the Caribbean-Dynamics of the Current Economic Cycle and Policy Challenges for Boosting Investment and Growth. *Herdon, VA: CEPAL*, Available at: https://repositorio.cepal.org/bitstream/handle/11362/42002/155/S1700699_ en.pdf.

CISCO Systems. (2016). Cisco Visual Networking Index: Global Mobile Data Traffic Forecast Update, 2016–2021 White Paper. *White Paper*. California, USA, 2016. Available at: https://www.cisco.com/c/en/us/solut ions/collateral/service-provider/visual-networking-index-vni/mobile-white-paper-c11-520862.html

Computer Weekly. (2017). GDPR: 10 Essential Facts You Need to Know. *E-Guide: CW+ Content. Computer Weekly*, Available at: https://www.computerweekly.com/ehandbook/GDPR-10-essential-facts-you-need-to-know

ECLAC. (2014). Cloud Computing in Latin America Current Situation and Policy Proposals. Santiago de Chile, Chile: Economic Commission for Latin America and the Caribbean (ECLAC), https://www.cepal.org

/en/publications/36740-cloud-computing-latin-america-current-situation-and-policy-proposals

E-Health Reporter Latin America, and HIMSS. (2018). Salud en la nube: Evolución hacia el nuevo paradigma en América Latina. *TIC/HIMSS/EVERIS*, www.ehealthreporter.com

GSMA. (2017). The Mobile Economy Latin America and the Caribbean. *London, UK: GSMA*, 2017,. Available at: https://www.gsmaintelligence.com/research/?file=e14ff2512ee244415366a89471bcd3e1&download.

Gutierrez, H. E. and Korn, D. (2014). Facilitando the Cloud: Data Protection Regulation as a Driver of National Competitiveness in Latin America. *University of Miami: Inter American Law Review*, Available at: https://inter-american-law-review.law.miami.edu/wp-content/uploads/2014/03/Facilitando-the-Cloud.pdf

Hamel, L., Jamie, F. and Mollyan, B. (2014). Kaiser Family Foundation/New York Times/CBS News Non-Employed Poll. Kaiser/New York Times/CBS News, Available at: https://www.kff.org/other/poll-finding/kaiser-family-foundationnew-york-timescbs-news-non-employed-poll/

Inter-American Development Bank (IDB). (2018). "Human Capital 2.0: The Future of Work in the Americas," 10. Lima, Peru. Available at: https://publications.iadb.org/handle/11319/8845

IPSOS, and Centre for International Governance. (2018). Majority (52%) Says They're More Concerned about Online Privacy than They Were a Year Ago - Factum. *Waterloo, Ontario Canada: IPSOS*, Available at: https://www.cigionline.org/sites/default/files/documents/CIGI-2018-Factum.pdf

Kirk, J. (2018a). Banco de Chile Loses $10 Million in SWIFT-Related Attack First, Attackers Distracted Bank, Using Buhtrap Malware to Cause Mayhem, Available at: https://www.bankinfosecurity.com/banco-de-chile-loses-10-million-in-swift-related-attack-a-11075.

Kirk, J. (2018b). Mexico Investigates Suspected Cyberattacks Against 5 Banks $20 Million in Potential Losses After Real-Tim Payment Connections Compromised. *Bank Info Security (blog),* https://www.bankinfosecurity.com/mexico-investigates-suspected-cyberattacks-against-banks-a-11008.

Liebenau, J., Patrik, K., Alexander, G., and Daniel, C. (2012). Modelling the Cloud: Employment Effects in Two Exemplary Sectors in the United States, the United Kingdom, Germany and Italy. *London, UK: LSE.*

Lus, L. (2018). Dealing with a Data Breach in Latin America: An Overview. Lexology, Available at: https://www.lexology.com/library/detail.aspx?g= b717c5ee-b3ff-42b1-92bf-a1c4e6405c95.

Magliona, C., Nicolas, Y., and Carlos, A. (2018). Data Protection and Privacy: Chile. *Santiago de Chile, Chile: Garcia Magliona y Cia Abogados,* Available at: https://gettingthedealthrough.com/area/52/jurisdicti on/3/data-protection-privacy-chile/

McKinsey and Company. (2018). A Future That Works: Automation, Employment and Productivity – January 2017. *Research Impact Insight. McKinsey Global Institute*, Available at: https://www.mckinsey.com/~/me dia /mckinsey/featured%20insights/Digital%20Disruption/Harnessing%20 automation%20for%20a%20future%20that%20works/MGI-A-future-that-works_Full-report.ashx.

Meltzer, J. P. and Marulanda, C. P. (2016). Digital Colombia: Maximising The Global Internet and Data for Sustainable and Inclusive Growth. *Global Economy and Development. Brookings Institute*, Available at: https://www.brookings.edu/wp-content/uploads/2016/10/global-20161020-digital-colombia.pdf

OECD, and Inter-American Development Bank. (2016). Broadband Policies for Latin America and the Caribbean: A Digital Economy Toolkit. Paris, France: OECD Publishing Paris, Available at: https://www.oecd-ilibrary.org/content/publication/9789264251823-en.

OECD. (2015). *Digital Government Review of Colombia – Towards a Citizen -Drive Public Sector. Ley Findings*. Paris, France: OECD, Available at: https://www.oecd.org/gov/digital-government/digital-gover nment-review-colombia-key-findings.pdf.

OECD/CEPAL/CAF. (2016). *Perspectivas Economicas de America Latina 2017 Juventud, Competencias and Emprendimiento.* Paris, France: OECD Publishing, Available at: https://www.oecd.org/dev/americas/E-book_LEO2017_SP.pdf

Paganini, P. (2018). International Clothing Chain C and A in Brazil Suffered a Data Breach. *Security Affairs (blog)*, Available at: https://securityaffairs. co/wordpress/75943/data-breach/ca-brazil-data-breach.html.

Panez, O., Paola, C. (2018). and Veronica, N. Data Protection and Cybersecurity Peru – Montezuma & Porto (Lima), Available at: https://latinlawyer.com/jurisdiction/1004859/peru

Perezalonso Eguia, Pablo. (2017). Data Protected – Mexico." Ritch Mueller, Heather y Nicolau S.C. Available at: https://www.linklaters.com/ en/insights/data-protected/data-protected—mexico.

Seco, A. and Muñoz, A. (2018). Panorama del uso de las tecnolo-
gias y soluciones digitales innovadoras en la politica y la gestion
fiscal. *Division de Gestion Fiscal. Sector de Instituciones para el
Desarrollo. Banco Interamericano de Desarrollo (BID)*, Available at:
https://publications.iadb.org/handle/123456789/6

Team Intricately. (2018). *Latin America Cloud Market Report (H1 2018)*.
San Francisco, USA: Intricately, Available at: https://www.intricately.
com/blog/latin-america-cloud-market-report-h1-2018

UN Secretary General. (2018). Strategy on New Technologies. NYC,
USA: United Nations, Available at: http://www.un.org/en/newtechnolog
ies/images/pdf/SGs-Strategy-on-New-Technologies.pdf

US International Trade Commission. (2017). "Global Digital Trade 1: Shifts
in U.S. Merchandise Trade 2014, 2014 Annual Report – Market Opportunities and Key Initiation Foreign Trade Restrictions." Investigation number?: 3322–561. Washington D.C, USA: United States International Trade
Commission, Available at: https://www.usitc.gov/publications/332/pub471
6_0.pdf

World Bank. (2016). The World Bank Annual Report. Washington,
D.C. ©World Bank. Available at: https://openknowledge.worldbank.org/
handle/10986/24985

2

Lessons from Brazil's National Computer Policy for a New Industrial Policy on Industry 4.0

Walter Shima, Pollyanna Rodrigues Gondin, Marcelo Castellano Lopes and Marcelo Vargas

Federal University of Paraná (UFPR), Brazil
waltershima@ufpr.br, pollygondin@gmail.com,
marcelolopescl@gmail.com, marcelo.vargas@unespar.edu.br

The goal of this study is to search Brazil's National Computer Policy for institutional elements articulating public policy articulation that would further promote the Industry 4.0 policy. The article presents a historical overview of the National Computer Policy from the 1970s and 1980s until its abandonment in the 1990s, as well as the articulation of policies promoted by the State. Data was collected using the RAIS platform. The data presents the evolution of the computer equipment and software development industries, which existed between 2006 and 2016. The results shows the presence of an industrial base that allows an initial investment and leverage for the development trajectory of Internet of Things (IoT), mainly in software development. The discussion in this chapter indicates that if the National Computer Policy had not been discontinued in the 1990s, Brazil could have reached a higher level of global competitiveness. Thus, a new National Industrial Policy 4.0 could transform Brazil into an important player in Industry 4.0, as this is a race that is yet to be dominated by potential global players. The arguments in the chapter is supported by 3 studies on Industry 4.0 made on Government agencies.

2.1 Introduction

The purpose of this study is to search Brazil's National Computer Policy of the 1970s and 1980s for institutional elements that articulate public policy that could support a National Policy in Industry 4.0. Internet of Things (IoT) is a critical component of Industry 4.0; hence, its national policy in the context of Industry 4.0 is prioritized in this discussion. In Brazil, the state played an important role in the articulation of the National Computer Policy. This policy is studied because it refers to an activity of paradigmatic nature such as the current IoT. In the 1970s and 1980s, a new technological paradigm emerged which had an impact on market entry. Here entry into the industry necessarily implied that the State acted as a driving force for stimulating, developing, and articulating private (national and multinational) capital. The Brazilian state acted in an articulate way to build institutions and define forms of action in order to create and consolidate the national industry. However, in Brazil this political articulation lost traction during the 1980s and 1990s due to missteps and policies that prioritized macroeconomic goals, which hindered the Brazilian computer industry from establishing itself as an internationally relevant *player*.

In this chapter, it is advocated that the emergence of IoT technologies opens new opportunities for a technological race towards industrial development through which Brazil can once again strive to become a major global competitor. This requires effective industrial policies. Obviously, the current scenario is different, and competitors from the "old" computer industry are rapidly diversifying their actions in the face of these new opportunities. However, as IoT implies the intensity of knowledge, it is possible to envision the development of solutions and actions focused on creativity. Brazil has a robust industrial base capable of developing IoT as will be discussed in section 2.4, based on data from RAIS/MTE. Therefore, Brazil has an "Information Technology (IT) industry" that expanded significantly between 2006 and 2016 in terms of number of establishments, employment, and training. This reflects its capacity to both absorb and produce knowledge.

As a starting point for an IoT industrial policy, Brazil's National Bank for Economic and Social Development (BNDES) recently published 27 reports detailing results of various research activities on this topic. The set of information gathered were based on the formation of expert panels, public surveys, interviews, and *workshops*. The research resulted from a public call for proposals won by a consortium led by McKinsey & Company, Fundação CPqD, and Pereira Neto Macedo Advogados. The study was extensive and provided a series of relevant data. However, as will be seen, the study does not present a conception for industrial development nor an industrial policy.

This chapter comprises of four more sections aside the introduction and the conclusion. The first section makes a case for why the State must act as the policy driver, as it is only from this perspective that it is historically possible to achieve development, as well as create the potential for a Public-private relationship, which is of fundamental importance. The second section provides a historical overview of the characteristics of Brazil's National Computer Policy, showing the role of the State in leading policies. In the third section, we analyze some data from what should be called here as the "IT industry." The fourth section presents an evaluation of the study carried out by BNDES. Finally, we present the conclusion.

2.2 The State as an Articulator and Policy Promoter

A country's industrial policy is a set of actions involving both the public and private sectors that aims to generate technological progress, increase industrial competitiveness, and contribute to the country's growth and development. This makes the industrial policy one of the key themes for the strengthening of an industry and other strategic economic sectors, contributing to the elaboration of a national development project. According to Kupfer (2003), it is up to the industrial policy to accelerate the productive transformation processes under which the market forces can operate. Thus, successful industrial policies do not go against the market; however, at the same time, they are not limited to complementing the market, as they have a greater scope.

Taking a similar approach, Krugman (1989) understands industrial policy as a governmental effort to promote economic growth through the development of strategic sectors. The idea of developing strategic sectors relates to key sectors and would have fundamental and broad effects for catching-up. Although there are approaches that are contrary to the development of industrial policies, international experience shows that all countries practice this type of policy, the only difference being the strategies and instruments used, the degree of activism, the priorities listed, and how the decision-making process is coordinated (Kupfer, 2003). In the same line of thought, Chang (2002) suggests that developed countries have historically used various types of industrial policies and are now "kicking the ladder" of those countries that perform the same types of policies in order to prevent them from compete at the same level of development.

Mazzucato (2013) presents an approach that is not only Keynesian, but also Schumpeterian. The author seeks to demystify the false idea that the freedom of the private sector will invigorate services. This widespread, false idea is based on the common-sense dichotomy between a revolutionary, dynamic, innovative, competitive private sector in opposition to a slow, backward, and bureaucratic public sector. According to the author, the State does not have a good marketing department and, therefore, it would be necessary to raise public awareness about the history of technology, placing emphasis on the relevant role played by the State. This highlights the key role played by the State in innovation processes, such as in the case of Silicon Valley, in which the State not only facilitates the knowledge construction and learning processes, but also acts in specific investments. In this respect, Mazzucato (2013) also states that the great radical innovations that have occurred, such as the railways, the internet phenomenon, and nanotechnology, would not be possible without the participation of the State – for example, in the provision of financing. The State is also relevant in cases of radical innovations, since these are risky and marked with a high degree of uncertainty. The State is therefore an important partner of the private sector, as it takes on the risks involved in the process of innovation.

At the same time, Mazzucato (2013) emphasizes that the role of the State is not limited to the management of Keynesian demand: it also acts as a businessman, risk taker, and market maker. The State assumes a central role in innovation systems, facilitating and creating conditions for the process of innovation. Considering that the diffusion of innovations is not linear, factors such as education, training, design, quality control, and effective demand are as important as the company's R&D (Research and Development) system. This, once again, reinforces the State's performance in innovation.

In addition, Mazzucato (2013) ensures that the State must be in command of the process of industrial development and the development of strategic areas, leading the process of industrialization. The defense of a more active State, according to the author, originated from a consensus among several countries that try to make for their backwardness regarding technology in relation to more developed economies. To do so, the State's actions must take local institutional specificities into account. The State should develop policies to support the enhancement of companies' technological capacity, including incentive schemes and support in the form of public service, in addition to funding research and development projects.

As an example, Mazzucato (2013) analyzes the Asian economies. According to the author, in the less industrialized countries, the State coordinated the whole industrialization process, leading and choosing sectors for investment and establishing barriers to foreign competition until companies in the target sectors were able to export their products. After this phase, the State was still important, as they did aid in finding new export markets. Once again referencing Chang's (2002) approach, during this phase, developed countries would be "kicking the ladder" so that developing countries would not benefit from it.

Another example of the role of the state in the industrialization process is its important contribution to Apple's success in the United States amid the free-market discourse adopted by the country (Mazzucato, 2013). According to the author, Steve Jobs' genius, entrepreneurial ability, and technical knowledge were relevant for such a success to be achieved. However, investment in technology by the State provided Apple with the foundation needed to launch its products. Prior to launching its most popular iOS platform products, Apple received considerable government support directly and indirectly through three major areas. The first area was the direct equity investment by the State during the initial stages of creation, risk, and growth (ibid). The second area was Apple being granted access to technology that had resulted from governmental research programs, military initiatives, public-law contracts, or had been developed by public research institutions (ibid). Finally, the third area was the creation of fiscal, trade, and technology policies by that sustained the company's innovation efforts (ibid).

Thus, an effective industrial policy presupposes a flexible State committed to the processes of change and that promotes an environment conducive to the exploitation of novelty. At the same time, such a policy must consider local and institutional specificities. Therefore, an industrial policy must be addressed with a focus on state action, competition, and technical progress. It is through state action in competitive environments that private companies are led to pursue innovation strategies (Gadelha, 2002).

In addition, state action must also consider the informatization of societies and economies, as well as the global dynamics, in order to implement national, regional, and local policies. Furthermore, the formation of geopolitical blocs, regional economic systems, and institutions such as the United Nations (UN) and the International Monetary Fund (IMF) shape the influence of the State and the development of public policies.

The following section will show that Brazil's National Computer Policy, in the 1970s and 1980s, permitted an attempt to proactively involve the State in the creation and development of a key industrial sector in alignment with the new economic growth dynamics.

2.3 Industrial Policy and Technological Trajectory of Brazil's Computer Industry

The idea and creation of a national computer industry in Brazil took place amid the global technological progress that followed World War II. In the 1970s, the Brazilian military government took interest in the computer sector driven by their concern about the costly importation of war equipment. This underscores the important role played by institutional factors and public policies in the development and selection of possible trajectories which combines the Brazilian computer industry with the interests of the country's armed forces.

According to Evans (1995), the vision of a local IT sector began with individuals who were convinced of the value of local computer manufacturers. The idea found support in the state apparatus, so that it transformed into policies and institutions designed to facilitate the local production. Brazilian public policies began in "incubators" or "greenhouses," which is an evidence of the role of the State as a "midwife" (promoting new business groups or inducing existing groups by way of protectionist policies, tariff barriers, subsidies, among other strategies) and its goal of protecting local entrepreneurs.

2.3.1 The Creation of the Computer Policy in the 1970s: The Active Role of the State and the Institutional Articulation

The first half of the 1970s (within the period that has become known as the "Brazilian economic miracle") produced a general confidence in an industrial transformation, mainly through the II National Development Plan, known by its Portuguese acronym II PND, which emphasized the improvement of Brazilian industrial potential and expanded the role played by local capital in this process. Despite the dominant free-market ideology, technocrats that were sympathetic to the nationalist industrial policy gained ground among policymakers. The early 1970s can be considered as a time when the Brazilian computer policy began to be outlined. As part of this effort, in March 1971, an agreement between the Navy, the Ministry of Planning, and Brazilian universities created the GTE/FUNTEC 111 (Special Working Group – Technological Fund) with an initial budget of US$ 2 million. This resulted in the construction of the G-10 minicomputer, with hardware developed by the Polytechnic School of the University of São Paulo (Poli-USP) and software developed by the Pontifical Catholic University of Rio de Janeiro (PUC-RJ) (Tigre, 1993).

In 1972, Coordinating Commission for Electronic Processing Activities (CAPRE) was created to implement government policies for the computer industry. The Ministry of Planning chaired CAPRE, and it had representatives from the Armed Forces, the Ministry of Finance, The National Bank for Economic Development (BNDE[1]), Federal Data Processing Service (SERPRO), Brazilian Institute of Geography and Statistics (IBGE), and the Office of Administrative Reform. One of CAPRES's guidelines, which later became the central axis of the National Computer Policy, was the *training of Brazilian industrial organizations in hardware development and manufacturing*. In an average term, the commission had its power expanded to such an extent that in February 1976 its composition changed, further providing it with powers to establish the guidelines of the National Computer Policy (Fajnzylber, 1993).

Around the same time, the creation of an instrument of state intervention was proposed with the purpose of stimulating the development of the country's electronics industry. This led to the creation, in 1972, of the Eletrônica Digital Brasileira Ltda., renamed DIGIBRÁ/SA after its nationalization, with the aim of enabling BNDE and other governmental institutions to participate in the establishment of national computer companies.

In 1973, amid the oil crisis, the government decided to act in order to solve problems such as the *deficit* in imports, and as such, the National Computer Policy was implemented to create local computer industries. In 1974, a Brazilian IT company called Computadores e Sistemas Brasileiros (COBRA) was founded with the equity participation of DIGIBRÁS, a UK-based electrical engineering and equipment firm Ferranti, and EE Equipamentos Eletrônicos (a national private company). The initial idea was to produce the Argus 700, a minicomputer designed for control and process applications. Although it was created in 1974, both partners and the technologies chosen

---

[1]Created in 1952 by the Law no 1.628, it was intended to be the body that formulates and enforces the national policy of economic development. The early 1980s were marked by the integration of social concerns into Brazil's development policy. Thus, in 1982, the Bank was renamed BNDES (The National Bank for Economic and Social Development).

were initially inadequate, so COBRA only became operational in 1977[2] (Fajnzylber, 1993).

Until 1975, multinational computer companies dominated the national market as multinational businesses increased at an accelerated rate. Thus, the number of installed computers grew and IBM *"despite maintaining its leadership in terms of the value of installed equipment, fell to third place in terms of the number of machines in operation, behind Burroughs and Olivetti"* (Fajnzylber, 1993).

In 1977, CDE (Economic Development Council) passed a resolution with the purpose of guiding government agencies to concede tax incentives and approve import orders for parts and components for the manufacture of computers in Brazil. This resolution had five points, namely:

- The degree of openness to absorb technology as well as the prioritization of companies that create and develop their products and production techniques based on national engineering;
- Nationalization indexes that prioritize companies with no permanent links with foreign suppliers that could hinder an effective nationalization of their products;
- The participation of companies or liberalization in the internal market, in order to avoid excessive concentration of the production;
- National ownership;
- and foreign currency balance, that prioritizes companies whose prospects are more favorable to the internal market (Fajnzylber, 1993).

In addition, the market reserve policy for the computer industry in Brazil was also implemented to control the imports and the actions of multinational subsidiaries in the country and to stimulate the creation of national minicomputer and peripheral equipment companies. According to Fajnzylber (1993), in 1977, CAPRE issued a favorable opinion to COBRA's project on the manufacturing the Sycor 400 and a to contrary opinion the manufacturing of the IBM system in the country. In the face of the refusal to IBM's

---

[2]"This year, after successive injections of funds made by BNDE (which, as the national partner, always showed its inability to contribute to the capitalization of the venture), a consortium consisted of thirteen national financial institutions gained ownership of 39% of COBRA's capital, while another 39% were undertaken by SERPRO, Banco do Brasil, and Caixa Econômica Federal: COBRA's equity went from US$4.4 million to US$30.8 million. In the same period, the company's product line was strengthened through a new technology transfer agreement referring to a general-purpose minicomputer (Sycor's 400 model), which, unlike Ferranti's model, had a significant market presence" (Fajnzylber, 1993, p. 8).

System/32, a national competition was launched for the minicomputer market, and three companies with 100% national capital and licensed technology were chosen. These companies were, Sharp/Inepar/Dataserv, with technology from Logabax, a French computers manufacturer; Edisa, with technology from Fujitsu, a Japanese technology company; and Labo, with technology from German computer company Nixdorf.

In 1979, there was an institutional reform, with CAPRE being replaced by the Special Secretariat for Informatics (SEI), linked to the National Security Council. This gave a new focus to the computer sector, affirming the supremacy of the military over BNDE and the Ministry of Planning. This led the computer sector to become a strategic sector (Rodrigues, 2004). Although there were other companies in the sector, COBRA was dominant in the market, defining the path to be followed by other national computer companies in the production of skilled workers and in the promotion of technological progress. COBRA did succeed in advancing technology and supporting the internal market (Tigre, 1993).

Regarding the domestic market, it is worth emphasizing that the national industry was focused on the production of minicomputers and peripheral equipment, while multinationals such as IBM, Burroughs, and Hewlett-Packard, produced larger computers (Alder, 1986). One of Brazilian computer industry's main achievements at the time was the development of a medium-sized computer, the COBRA 530 model, and the CYSNE 2.0 operating system, created along the lines of the Microsoft Disk Operating System (MS-DOS). The success of Brazil's computer industry was related to two forms of state action: the investments made in the state-owned COBRA, which guaranteed an R&D program, and the protection of the internal computer market with the establishment of a contract between society and government based on the idea of national sovereignty.

2.3.2 Computer Policy in the 1980s: The Attempt of Intensifying the Industrial Policy and Strategic Errors About the Technological Trajectory

The "informatics law" was sanctioned in October 1984. This policy consolidated the national computer policy that was being implemented up to that moment. The goal of this policy was *"capacity-building in national computer activities with a view of benefiting the social, cultural, political, technological, and economic development of Brazilian society" (article 2 of Law no 7.232/84)* (Fajnzylber, 1993). Although it followed the guidelines of the policy been implemented since CAPRE, the new law proposed changes in the institutional framework. Therefore, normative actions became the responsibility of National Council on Informatics and Automation (CONIN). One of CONIN's functions was to propose the National Plan for Informatics and Automation (PLANIN) every three years.

The I PLANIN was approved through Law no 7.463, enacted in April 17, 1986. It essentially established criteria for the execution of the national computer policy over a three-year period. As of December 1985, the incentives provided for by law were regulated. Among these incentives were: exemption or reduction of the import tax, exemption from the export tax, exemption or reduction of the tax on Industrialized Products (IPI), exemption or reduction on Tax on Financial Operations (IOF), among others. As a compensation for the granting of incentives, *the mandatory percentage of R&D investments of commercialization revenues obtained by the beneficiary computer companies was set at 10%* (Fajnzylber, 1993).

The microelectronics sub-industrial sector benefited more from the tax incentives, corresponding to 93% of the 50 companies were granted these incentives between 1986 and 1987. However, the appropriate incentives were lower than those allowed (Fajnzylber, 1993). However, the author also states that the government's financing of the computer industry and purchasing of goods produced by domestic companies was late and small. Furthermore, Brazil did not have or develop an expressive science and technology infrastructure.

Thus, from the second half of the 1980s, the national interest was on the wane since the national industry could not keep up with the microelectronics revolution that developed countries around the world were experiencing. The excessive protectionism that underpinned all incentive mechanisms precluded national companies from achieving the level of competitiveness and dynamic technological development of central countries. The policy lacked a vision of competitive integration, which had been discussed since the early 1980s. As an additional reflection of this situation, COBRA, the company that defined the entire computer sector, was experiencing internal problems related to the finances and lack of incentives for employees.

Another issue that deserves to be highlighted refers to the new nature of Brazilian public policies at that point in history, when developmentalism gave way to a neoliberal vision of the state, changing the course of development of Brazil's computer industry.

2.3.3 The End of the Computer Policy from the 1990s: Neoliberalism and the Dismantling of the State

In 1991, the then Brazilian President Fernando Collor ended the Computer Policy, putting an end to the institutional apparatus and the instruments of politics. Thus, the protectionist approach taken by the country in the 1970s to stimulate its technological development was abandoned in favor of trade liberalization initiated in the 1990s. This act reveals how fragile the Brazilian computer policy was in relation to policies that worked in developed countries (Tigre, 1993).

According to Tigre (1993), from then on, the survival of the Brazilian computer companies depended on a reformulation of their strategies, the seeking of new partners, the redirection of production to other sectors, or even the abandonment of the market. Consequently, the national companies that remained in the computer market reduced their activities and transferred them to new niches. At the same time, companies started to acquire foreign computer-related technology and products for the domestic assembly of computers, reducing Brazilian companies' R&D expenses to about 70%.

Therefore, operating in the personal computer market has become increasingly difficult for domestic companies, mainly due to the intense pace of innovation and the significant cost reduction decided by multinational companies both in Brazil and abroad. Multinationals began to dictate global technology standards through proprietary operating systems, with which it would be impossible to compete or establish some competitive capacity on an equal footing. The weak national industry should seek limited market alternatives and, consequently, of technology. To face this, it would be necessary to have a technological articulation and the production of networks by the manufacturers themselves, thus finding the vocations in the informatics sector that will form a national and/or sectorial system of innovation. Something that, though timidly, is still present in the irrelevant industrial policy of the country.

2.4 Perspectives of a Vertical Industrial Policy with Focus on IOT

This survey about the recent economic history of the industrial policy in Brazil shows that the decision to develop a national computer industry depended on a proactive Government decision. This led to a national articulation between several institutions within the public sector and the creation of institutional/legal conditions of a political nature and consequently a power structure among the many areas involved. In other words, the possibility for developing a national industry would not be the result of eventual policies or just competitive free market dynamics, but of an explicit political, economic and legal/institutional action. It is true that abandonment or due importance were not given to the sector in the 1980s due to the pursuit of the exclusive macroeconomic adjustment aimed at combating inflation. Moreover, the official abandonment of the industrial policy in the early 1990s definitely condemned Brazil's entry into the global competition of the new technological paradigm. If, on the other hand, the informatics policy had followed a sequence, it would have been possible to become a major player in the production of Information Technology (IT), just as South Korea. In the 1970s and 1980s, South Korea was in a lower position than Brazil in terms of technological development, and by the end of the 1990s it had become an international *player* in the niche of televisions, video monitors and other segments of IT. In a simple analogy, Brazil abandoned the global IT race in the middle of the road, while South Korea did not. The latter did not necessarily reach the top of the podium, but they got to the top and now occupies an important position.

Globally, in the 70s/80s the IT industry was not yet consolidated. There were no established important *players* already and each *one* would adopt their own business, market and technology strategy. There were several proprietary operating system standards, auto-fabricated processors, monitors of several kinds, among others. Every competitor sought some competitive advantage regarding price and friendliness of the equipment. The race unraveled in terms of both *hardware* and *software*, and the big competitive issue was to become the market standard and thus create network externalities. Therefore, in the 1990s, given the various strategies of the firms with strong support from the national states, the Intel-Microsoft-IBM standard was consolidated and the other competitors had to rearrange themselves in the different possible niches within this standard. At the same time that many simply left the market, such as Remington-Rand, Radio Shack, Olivetti, MicroPro International, Sorcim, Borland, among others.

The *players* in the current IT industry (which we can say is becoming "old") are already consolidated and the entry as a global *player* of any firm in any of its niches is virtually blocked. In addition, this input would imply doing more of the same, only with incremental innovations. National States already have a clear understanding that industrial policies facilitating entry into this industry will be innocuous and that it is an industry whose leaders are already consolidated.

One of the elements that the current dynamics of technological development has radicalized is the development of informatics, from the incorporation of processors to all objects that can communicate through the Internet using a standard communication protocol. It is radical because the processors are becoming smaller and with greater capacity of communication at decreasing costs. Since everything communicates with everything on a single network through a standard protocol (in which global *players* also compete to enforce their standard), the immediate consequence is the creation of a growing database. Processors along with sensors embedded in objects can predict and anticipate a series of events in all areas of human life. It will be more and more possible to generate a database for all types of activities and control of human actions and objects. This is the IoT which is part of Industry 4.0. Industry 4.0 comprises of a set of converging technologies of the physical, computer and biological world (additive manufacturing, artificial intelligence, IoT, synthetic biology and cyber-physical systems). Figure 2.1 illustrates Industry 4.0.

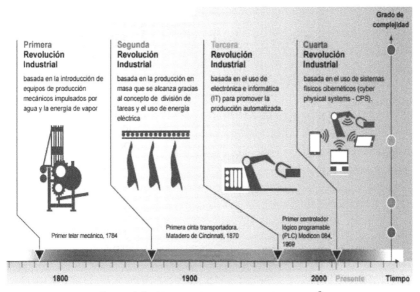

Figure 2.1   From Industry 1.0 to Industry 4.0.[3]

---

[3]http://grupofranja.com/index.php/oftalmica/item/1763-de-la-industria-1-0-a-la-4-0. Accessed on 04/20/18.

According to Hermann et al. (2015) quoted by Furtado et al. (2017):

"Industry 4.0 is a collective term for technologies and concepts of the value chain organization. Within Intelligent and Modular Manufacturing Factories 4.0, cyber-physical systems (CPS) monitor processes, create a virtual copy of reality, and make decentralized decisions. Through the Internet of Things (IoT), the CPS communicate and cooperate with each other and with human beings in real time, and through the Internet of the Services (IoS) internal and external organizational services used by participants in this value chain are offered".

It is possible, therefore, to realize that it is an industry that is still in its infancy and, consequently, there are no leading *players* in the molds established in the "old" computer industry. The possibilities for developments within Industry 4.0 are not entirely appropriate, even though the "old ones" can diversify into this new industry and are already pursuing heavy enforcement strategies for domination and standard. These possibilities are more closely linked to the creativity in developing solutions and ideas that lead to automation and greater human-machine interaction. Consequently, the mechanisms of barriers to entry are not fully defined. Given the high degree of uncertainty, competitors are still setting up themselves based on their technological, market and regulatory expectations[4].

Therefore, just as in the 1970s and 1980s, a window of opportunity for industrial development with informatics that was not much utilized by the industrial policy in Brazil has opened up. Industry 4.0 is that new open window with broad possibilities of entry. It is a new race in which the conditions favor Brazil to participation on equal footing and in considerable competitive conditions. This is because the characteristic elements of this new race are linked to the ones aforementioned. The industrial base is important in its applicability and development, but the relevant aspect is that the possible segments to turn to the development of 4.0 have expanded significantly in the country. In that sequence, a supposed industrial base will be presented, which is available in Brazil and important for the development of IoT in particular.

2.4.1 Evolution of CNAE[5] Segments as Possible IOT Developers

The idea in this section is to analyze the behavior of the number of establishments, employment and qualification from the data of RAIS of Division 26

---

[4]Given the novelty feature, the regulatory processes by the National States are also still uncertain. It is not even clear what, much less how, this regulation should be.

[5]National Classification of Economic Activities.

(Manufacture of Computer Equipment, Electronic and Optical Products), and Classes 6201–5; 6202–3; 6203–1 from Division 62 (Activities of Information Technology Services). These are the classifications assumed here as those segments with broad potential to become or develop into IoT activities. The guiding questions here are what is the scale of these activities in Brazil? Is there an important industrial base?

According to Tables 2.1 and 2.2, in the periods 2006/16 Micro and Small enterprises accounted for more than 90% of the total establishment at the beginning and end of the period. In Division 26 there is relative stability in the Compound Annual Growth Rate (CAGR[6]) of the number of establishments considering the low and even negative values of the CAGR by the size of companies.

On the other hand, in the classes (Table 2.2) studied; there is an important expansion in the number of establishments that can be more or less correlated to the size. That is, the number of companies grows more as the size increases. As an example, the CAGR of the micro companies (8.83%) compared to the large ones (13.54%). The synthetic idea from these data is that more activities related to knowledge and creativity (software) tend to grow than production

Table 2.1 Number of Establishments in Division 26 – Manufacture of Computer, Electronic and Optical Equipment – 2006–2016

Size	2006	%	2016	%	CAGR
Micro	2,160	71.19%	2,532	74.89%	1.46%
Small	616	20.30%	636	18.81%	0.29%
Medium	211	6.95%	181	5.35%	−1.38%
Large	47	1.55%	32	0.95%	−3.43%
Total	**3,034**	**100.00%**	**3,381**	**100.00%**	**0.99%**

Source: Elaborated by authors based on RAIS-MTE.

Table 2.2 Number of Establishments in Classes 6201-5; 6202-3; 6203-1 – Software Development – 2006–2016

Size	2006	%	2016	%	CAGR
Micro	2,419	75.31%	6,138	71.58%	8.83%
Small	608	18.93%	1,818	21.20%	10.47%
Medium	111	3.46%	320	3.73%	10.10%
Large	74	2.30%	299	3.49%	13.54%
Total	**3,212**	**100.00%**	**8,575**	**100.00%**	**9.34%**

Source: Elaborated by authors based on RAIS-MTE.

---

[6]$TCAC = \left[\frac{valor\ final}{valor\ inicial} \right]^{\frac{1}{n}} - 1$; Where n = number of periods.

activities that do not necessarily have the same degree of knowledge require-ment. Production activities, as noted in Table 2.1, are declining activities in terms of number of establishments.

Likewise in Table 2.1, the CAGR of employment in the same Division 26 also remained relatively stable, and also in the same way that in Table 2.2 employment grew significantly in the Classes.

Considering Table 2.1, in which there was a slight evolution (0.99%) in the number of establishments in Division 26, it can be seen in Table 2.3 that this evolution did not affect the number of active links, since the variation was negative (–1.08%). This result is best seen in Table 2.5 below, in which the level of education was negative where there is lower education and positive where there is higher education, showing a change in the profile of the workforce. In Tables 2.3 and 2.4, in 2006, employment in Division 26 was 2.5 times greater than employment in the Classes and given the total CAGR of 10.89% (in Classes) in the period, employment in classes exceeded Division 26. Another element that shows that the most intensive activities in knowledge are the ones that develop most in the IT industry.

Table 2.3   Number of Active Links in Division 26 – Manufacture of Computer, Electronic and Optical Equipment – 2006–2016

Size	2006	%	2016	%	CAGR
Micro	12,988	9.24%	13,886	11.14%	0.61%
Small	26,875	19.13%	27,669	22.19%	0.27%
Medium	44,535	31.70%	38,907	31.20%	–1.22%
Large	56,112	39.93%	44,241	35.48%	–2.14%
Total	**140,510**	**100.00%**	**124,703**	**100.00%**	**–1.08%**

Source: Elaborated by authors based on RAIS-MTE.

Table 2.4   Number of Active Links in Classes 6201-5; 6202-3; 6203-1 – Software Develop-ment – 2006–2016

Size	2006	%	2016	%	CAGR
Micro	6,295	11.42%	17,620	10.26%	9.81%
Small	12,743	23.13%	37,371	21.76%	10.28%
Medium	7,549	13.70%	22,239	12.95%	10.32%
Large	28,514	51.75%	94,477	55.02%	11.51%
Total	**55,101**	**100.00%**	**171,707**	**100.00%**	**10.89%**

Source: Elaborated by authors based on RAIS-MTE.

Table 2.5 Degree of Education in Division 26 – Manufacture of Computer Equipment, Electronic and Optical Products – 2006–2016

Schooling After 2005	2006	%	2016	%	CAGR
Illiterate	123	0.09%	34	0.03%	−11.03%
Middle School (Complete and Incomplete)	20,099	14.30%	7,378	5.92%	−8.71%
High school (Complete and Incomplete)	94,844	67.50%	86,929	69.71%	–0.79%
Higher Education (Complete and Incomplete)	25,305	18.01%	29,800	23.90%	1.50%
Master's degree	118	0.08%	485	0.39%	13.71%
Doctorate degree	21	0.01%	77	0.06%	12.54%
Total	**140,510**	**100.00%**	**124,703**	**100.00%**	**−1.08%**

Source: Elaborated by authors based on RAIS-MTE.

Another relevant element that demonstrates the expansion of the IT Industry in the country is the improvement in the qualification of the workforce. This reflects the ability to absorb, learn and develop R&D. As seen in Table 2.5, it is possible to notice the decrease in the CAGR of the less qualified strata and increases in the CAGR of the most qualified stratum. Although the middle stratum (complete and incomplete) is still the largest employer in absolute terms, there was a significant expansion of Master's and Doctorate degrees.

At the same time, following the trend of the previous tables, Table 2.6 shows that the classes had an expansion in education bigger than Division 26. Here again, the need for greater qualification is evident, since the bulk of the workforce is from the higher education (complete and incomplete) stratum and employs in absolute terms more than the high school stratum of Division 26 (135,187 against 86,929). There was a greater expansion of this activity in relation to the higher Computer Supported Cooperative work (CSCW) in all strata and mainly in the higher, master's and doctorate strata vis-à-vis the same strata in Division 26. In other words, Classes qualify more in terms of training than Division 26. This fact is logic insofar, as already said, it is an activity of greater intensity of knowledge and, in this case, it is improving for its production, which increases its absorptive capacity.

In summary, the data shows that in the IT industry in Brazil the activities of greater intensity of knowledge, which are the classes, evolved more intensely than the activities of Division 26 that refer to the industrial production of equipment in general. This implies lower requirements for the knowledge intensity. In this sense, the expansion of the classes is a sign of

Table 2.6   Degree of Schooling in Classes 6201-5; 6202-3; 6203-1 – Software Development – 2006–2016

Schooling After 2005	2006	%	2016	%	CAGR
Illiterate	15	0.03%	15	0.01%	0.00%
Middle School (Complete and Incomplete)	1,821	3.30%	1,926	1.12%	0.51%
High school (Complete and Incomplete)	17,585	31.91%	32,950	19.19%	5.87%
Higher Education (Complete and Incomplete)	35,552	64.52%	135,187	78.73%	12.91%
Master's degree	97	0.18%	1,438	0.84%	27.78%
Doctorate degree	31	0.06%	191	0.11%	17.97%
Total	**55,101**	**100.00%**	**171,707**	**100.00%**	**10.89%**

Source: Elaborated by authors based on RAIS-MTE.

the availability of an important "industrial" basis, or increasing ability to develop IOT activities. A vertical industrial policy in certain potential growth activities of IOT would have broad field for development.

2.5 The Nature of the First Steps of Industrial Policy of Industry 4.0

In 2016, Brazilian Development Bank (BNDES) launched the Public Call BNDES/FEP Prospecção nº 01/2016 – Internet-of-Things (IoT), whose objective was to "select proposals to obtain non-refundable financial support for independent technical studies with the objective of conducting a diagnosis and proposing public policies regarding the Internet-of-Things (IoT) theme." McKinsey & Company, Fundação CPqD and Pereira Neto|Macedo Advogados formed a consortium to win the bid[7]. Recently, the results of the consortium's studies made up of 27 reports detailing their research activities and results were published. Their methodology was basically based on the formation of expert panels, public consultations, interviews and *workshops*. The "demand-pull" focus of the research stands out. The summary sentence of the study shows the aspiration of Brazil in IoT. The sentence was: "*....Accelerate the implementation of the Internet of Things as an instrument of sustainable development of Brazilian society, capable of increasing the competitiveness of the economy, strengthening national productive chains, and promoting improvement of the quality of life....*[7]". This statement reflects

---

[7] Available at: https://www.bndes.gov.br/wps/portal/site/home/conhecimento/pesquisaedados/estudos/estudo-internet-das-coisas-IoT.

the fact that the study does not have a theoretical backbone and presents itself more as an information management system for generating guidelines. Thus, a diffuse conception of demand is predominant, which may indicate different paths from particular interests (depending on the worldview of respondents) instead of a structured conception of development. The demand-pull perspective considers that the market would be determinant in defining the direction of the innovative process as if it knew in advance the possible technological trajectories and rationally chose the best alternative. It is an approach incompatible with the historiography of innovation (see Rosenberg, 1983 and Dosi, 1982).

From some consultation activities and from the panels came the perception that, it was more important to survey the supply capacity than the demand capacity. One then wonders, what is the potential for IoT development in the country? From their finding, in product 3^8 there is a survey of such potential. The study starts from a technical division on the layers of IoT, namely: Devices, Networks, Service Support, and Applications and Security. Within each layer, there is an extensive set of sub layers, which would be the specific productive activities in IoT. For each layer, the study points out that there are established (incumbent) firms and points out incomers, which indicates that it is a race that has not yet been defined. However, there is an urgency, because the study shows that there are already a considerable number of multinationals operating in the Brazilian market, and that, judging by what was mentioned in the consultations, are fast becoming the leaders of this activity. From the incumbent firms and the mapped incomers, it is important to define their real activities and the effective possibilities of generating externalities for the national economy. The question is: to what extent are such activities capable of creating internal capabilities and generating industry competitiveness, as a whole, from the effects of *spillover*?

There is a whole plan of action and subsidies for the National IoT Plan. However, it is important to think about the significance of the launch and the basis of this Plan. As seen in the Information Technology Policy, as an example, BNDES historically played a central role in the elaboration and execution of industrial policy in Brazil. More recently, it has become one of the 10 largest investment banks in the world. Its ability to act in the management

---

[8] Analysis of supply and demand – Report – Analysis of the Offering: https://www.bndes. gov.br/wps/wcm/connect/site/6c597bfe-b92d-4084-ab07-5498e1ae2445/producto-3-analysis-de-supply-and-demand-report-partial-analysis-of-demand.pdf? MOD = AJPERES & CVID = lSZJkHO.

and direction of public and private investment is something that cannot be ignored. BNDES has also created a body of highly technical-competent staff in a wide range of areas, to reflect on national development, and with proper training. In this sense, it is curious that a study of fundamental importance for national development was led by a multinational private consultancy, allied to a law firm, with BNDES approval through the public call. There are no plausible reasons to undertake studies on development from this perspective, considering that these are public policies and that there are expertise within BNDES capable of creating proactive structures in the institutional articulation within the State. This fact led, as aforementioned, to a baseless study of theoretical thinking compatible with industrial policy, unable to indicate in its guidelines the purpose in terms of national development. In the end, the result holds a lot of relevant information, without clarity as to the objectives in terms of this national development.

Another element that does not provide clarity on the idea of an industrial policy is the view about the organizations of the economy sectors. Instead of adopting the concepts in each sector, system or complex that could find compatibility with those in the databases of CNAE 2.0 and from there think of national versus foreign firms and/or leaders versus marginal firms in order to formulate the policy with priorities, the study adopted the idea of "Environments". The concept of "Environments", which does not seem to have reference in the literature of industrial organization, covers users, suppliers and those who demand IoT, in what could be an industrial complex. However, it does not differentiate the nature of the offer; the studies only understands that it is important that the development of IoT (regardless of how the market structure is organized) "improve the lives of the 4 Environments that have been divided into Health, City, Rural and Industry."

In summary, what is perceived is that these are some general guidelines without a clear view on development and industrial policy. This is obvious, since the processors of the study, being a consulting firm, are more focused on an administration perspective, without understanding the true role of the state. Industrial policy, after all, presupposes a perspective of development, which is not found in the study. This all stems from the current Government's view that the State does not have to act in targeting the economy and, therefore, much less have any targeting of credit.

Two other government initiatives also isolated and disjointed from a national policy are the Brazilian Industry Agenda 4.0 of the Ministry of Industry. Trade and Services (MDIC) and the "CT&I Plan for Advanced Manufacturing in Brazil" of the Ministry of Science, Technology, Innovation and Telecommunications (MCTIC, 2017). The first initiative comes down to data, assessments and an initial proposal on the government[9] website page about Industry 4.0. Despite the very relevant and comprehensive information, it is only an informative portal of the initial possibilities and indications of possible directions of an eventual industrial policy for IoT. The portal indicates the formation of the Industry Working Group 4.0 that had the role of only elaborating this proposal of a possible policy. In a broader, more articulate and politically relevant perspective included in the Government's, this group could be a relevant body for unifying, managing and monitoring a national policy on Industry 4.0. The second is also more specific (with that of Brazilian National Bank that refers to IoT, only), since it refers to advanced manufacturing. This plan has as its starting point the document "Prospects of Brazilian Experts on Advanced Manufacturing in Brazil: an account of workshops held in seven Brazilian capitals in contrast to international experiences" prepared by a task force composed of MDIC and MCTIC technicians. It was a survey conducted in workshops in several states with the participation of businesspersons, specialists and technicians from the Ministries. Again, the focus is "demand-pull". However, the Plan also presents an important volume of relevant technical information and makes a number of institutional arrangements necessary for the implementation of an advanced manufacturing policy. From the previous one, what is important is not the discussion of the content of the studies (which do present relevant content), but rather the fact that there are three policy perspectives in the scope of Industry 4.0 that are disjointed and without prior knowledge among each other, since they are not referenced in the publications. Ultimately, there is incoordination and several possibilities that indicate absence of a public policy in practice.

2.6 Conclusion

This article showed that the State has historically had a central role in defining the guidelines and direction of economic development. Mazzucato's (2013) reference is central to retrieving the Keynesian state's view on demand management, risk taker and market maker. The State assumes a central role in innovation systems, facilitating and providing conditions for the innovative process. The risk assumed by the State is related to the fact that the diffusion

---

[9]http://www.industria40.gov.br/

of innovations is not linear, the roles of education, training, design, quality control and effective demand assume positions as important as the R&D system internal to the company. Moreover, in this the state takes these activities for itself.

As it can be concluded from these references, the State in Brazil would have a more central role in directing development since, according to Chang (2002), the developed countries would be acting to "kick away the ladder". That is, they "climbed the ladder" from the state and industrial policy, achieved development and now preach the Minimum State and the efficiency of the private sector to developing countries. Apple's success story is never told as a case in which the state played a central role.

In this line of strong action of the State, it is argued in the article that in the years 70/80 the Brazilian State had a strong action in the definition and conduction of the Informatics Policy. The central point is that there is a broad institutional articulation within the framework of the State apparatus. The Informatics Policy has its beginning from a contract between the Navy, the Ministry of Planning and universities as well as the creation of the GTE/FUNTEC 111 project. CAPRE was also created, being responsible for implementing government policies for the information technology sector. It was an institutional articulation involving the Ministry of Planning, Armed Forces, Ministry of Finance, Brazilian Development Bank (BNDES), Federal Data Processing Service (SERPRO), Brazilian Institute of Geography and Statistics (IBGE) and Office of Administrative Reform. One of the guidelines of this Commission, which later became the central axis of the National Information Technology Policy, was the *"qualification of Brazilian industrial organizations in the development and manufacture of equipment (hardware)"* (Fajnzylber, 1993). From these actions there was a relative profusion of domestic production with partnerships and licensing of foreign firms: Sybar 400 from COBRA, Sharp/Inepar/Dataserv with French Logabax technology, Edisa with Japanese Fujitsu technology and Labo with German Nixdorf technology. However, in 1991, the then President Fernando Collor, ended the Information Technology Policy, putting an end to the institutional apparatuses and the instruments of politics.

Nevertheless, throughout the 1990s and 2000s with all the growing mechanism of dismantling the state, privatizations and end of public policies, the Brazilian economy built an "IT industry". The RAIS/MTE data show an important expansion in the area of hardware and software manufacturing in 2006/16, in terms of expansion of the number of establishments, employment and training. The activities most related to the production of software had a more expressive expansion. This indicates a strong potential of these activities if they return to the production of IoT activities.

The first steps towards an Industry 4.0 policy demonstrate many weaknesses in our organization. Chronologically, the MDIC/MCTIC gave the first signs of contemplating a policy in November/16 with a workshop titled "Brazilian Expert Perspectives on Advanced Manufacturing in Brazil: an account of workshops held in seven Brazilian capitals in contrast to international experiences." Consequently, in December 2017, MCTIC published the "CT&I Plan for Advanced Manufacturing in Brazil – ProFuturo – Production of the Future". In 2017, the BNDES also launched studies on IoT. These studies, in spite of the fact that they bring important technical elements to the potential for developing IoT and advanced manufacturing in the country, it is more an information management exercise for the definition of technical guidelines. However, as mentioned previously, industrial policy implies institutional organization and, for this, it is necessary to define the political nature of the State. In this way, there is an incompatibility between the origin of the study and the meaning of an industrial policy. In other words, industrial policy means the articulation and creation of power structures within the state. These structures gain power in the sense of knowledge and in the capacity to establish the direction of politics. The State in this case is clearly not neutral. It takes a direction, once a private consortium carries out the study. There are no slightest condition for this private consortium to indicate the mechanisms of reorientation to the State. At the same time, the study is a rupture with the previous pattern of policymaking. This rupture involves the perspective that it does not have a theoretical foundation that discusses industrial policy and development, as well as because it seeks the definition of industrial policy in the private sector. Since, the various instances of the Brazilian public sector and BNDES, for example, are equipped with extensive technical competences to think and elaborate public policy subsidies. While the European Union, Japan and the United States are actively working on the development of Industry 4.0, articulating the State with the various industrial sectors in the elaboration of specific policies, in Brazil we have only 3 uncoordinated indications of possible policies. It demonstrate the lack of concern by the government to lead an action that may be the resumption of economic development. Thinking about the analogy made at the beginning of this article, we are under the risk of losing or not even starting this new development race, as what happened in the race of the national computer industry.

References

Alder, E. (1986). Ideological "Guerrillas" and the Quest for Technological Autonomy: Brazil's Domestic Computer Industry. *Journal Storage, source International Organization* Vol. 40 no 3, 1986. p. 673–705.

Chang, H. J. (1994). *The Political Economy of Industrial Policy*. New York: St. Martin's Press.

Chang, H. J. (2002). *Kicking away the ladder: development strategy in historical perspective*. Londres: Anthem Press. (PORTUGUES).

Dosi, G. (1982). Technological paradigms and technological trajectories: A suggested interpretation of the determinants and directions of technical change. *Research Policy*, 11(3), 147–162. https://doi.org/10.1016/0048-7333(82)90016-6

Evans, P. (1995). *Embedded Autonomy: States and Industrial Transformation*. Princeton, NJ, Princeton University Press.

Fajnzylber, P. (1993). *A Capacitação Tecnológica na Indústria Brasileira de Computadores e Periféricos: do Suporte Governamental à Dinâmica do Mercado*. Dissertação de Mestrado UNICAMP. Disponível em: http://repositorio.unicamp.br/jspui/bitstream/REPOSIP/286491/1/Fajnzyl ber_Pablo_M.pdf Acesso em: 28/03/2018.

Furtado (Responsável), J., Pinheiro, H., Urias, E. and Muñoz, D. (2017). Indústria 4.0: a quarta revolução industrial e os desafios para a indústria e para o desenvolvimento brasileiro. Julho/2017. http://www.iedi.org.br/media/site/artigos/20170721_iedi_industria_4_0.pdf. Acesso 13/04/18

Gadelha, C. A. G. (2002). *Estado E Inovação: Uma Perspectiva Evolucionista*. Disponível em: <http://www.ie.ufrj.br/images/pesquisa/publica coes/rec/REC%206/REC_6.2_04_Estado_e_inovacao_uma_perspectiva_evo lucionista.pdf> Acesso em: 02/09/2013.

Krugman, P. R. (1989). *Industrial organization and international trade*. In: Schmalensee, R.; Willig, R. (Eds.). Handbook of industrial organization. New York: Elsevier.

Kupfer, D. (2003). Política industrial. *Econômica*, Rio de Janeiro, v. 5, n. 2, p. 281–298.

Mazzucato, M. (2013). *The Entrepreneural State: Debuking Public vs. Private Sector Myths*. Editora Anthem Press.

MCTIC. (2017). Plano de CT&I para Manufatura Avançada no Brasil – Pro-Futuro – Produção do Futuro. Brasília – Dezembro/2017. Disponível em.

https://www.mctic.gov.br/mctic/export/sites/institucional/tecnologia/tecno logias_convergentes/arquivos/Cartilha-Plano-de-CTI_WEB.pdf

MDIC/MCTIC. (2016). Perspectivas de Especialistas Brasileiros sobrea Manufatura Avançada no Brasil: um relato de workshops realizados em sete capitais brasileiras em contraste com as experiências internacionais. Novembro/
2016. Disponível em http://www.abinee.org.br/informac/arquivos/maavm dic.pdf

Rodrigues, H. S., Rastro de Cobra. Rio de Janeiro. (2004). Caio Domingos & Associados Publicidade LTDA.

Rosenberg, N. (1983). *Inside the Black Box*. Cambridge University Press. 1983. https://doi.org/10.1017/CBO9780511611940.

Tigre, P. B. (1993). Ciência e Tecnologia no Brasil: Uma Nova Política para um Mundo Global liberalização e capacitação tecnológica: o caso da informática pós-reserva de mercado no Brasil. Instituto de Economia Universidade Federal do Rio de Janeiro. Rio de Janeiro.

Additional Resources

BNDES (2016). https://www.bndes.gov.br/wps/portal/site/home/conhecim ento/pesquisaedados/estudos/estudo-internet-das-coisas-IoT

_____ http://www.valor.com.br/empresas/5400829/plano-de-%3Finternet-das-coisas%3F-sera-lancado-em-abril-diz-governo

IBGE – Pesquisa Industrial Anual – Empresa https://www.ibge.gov.br/estatis ticas-novoportal/economicas/industria

Ministério da Industria, Comércio e Serviços. http://www.industria4

RAIS – Relação Anual de Informações Sociais http://bi.mte.gov.br

3

What Developing Countries Can Learn From The EU's GDPR

Roslyn Layton

Center for Communication, Media and Information Technologies,
Aalborg University, Denmark
roslyn@layton.dk

This chapter explores the theories and outcomes to date of the European Union's General Data Protection Regulation and preliminary takeaways for developing countries. Two relevant areas for developing countries are the impacts of regulation to small and medium-sized enterprises (SMEs) and the cost of compliance. The chapter concludes with recommendations for policymakers in developing countries.

3.1 Introduction

The purpose of the GDPR is to regulate the processing of personal data. The protection of persons in the processing of such data is deemed a fundamental EU right (Regulation (EU) 2016/679). Specifically, the GDPR is legislation from the European Parliament composed of 173 recitals which cover 45 specific regulations on data processing, 43 conditions of applicability, 35 bureaucratic obligations for EU member states, 17 enumerated rights, eleven administrative clarifications, nine policy assertions, five enumerated penalties, and two technological allowances. The legislation applies to topics including Rights of Rectification and Erasure, Restriction of Processing, Objection to Direct Marketing, and requirements for businesses to perform risk assessments, hire data protection officers, and conduct international data transfers.

The European Commission's GDPR website claims that the goals of the regulation are to give users more control of their data and to make business "benefit from a level playing field" (European Commission, 2018). Ostensibly the idea is that the regulation would reduce the market power of the American giants which dominate the European internet, e.g. Google, Facebook, and Amazon. A related goal is to facilitate growth for small and medium sized companies (SMEs) to compete. While not explicit, it is conceivable that policymakers desire to promote European firms over foreign ones. This chapter explores two areas of importance for developing countries, notably the impact to small and medium sized enterprises (SMEs) and the cost of compliance. Learning from the experience of European SMEs may be instructive for developing countries because SMEs are common in developing countries and are described as nucleus of economic activities, a major source of employment, key contributors to economic growth and are engaged in poverty eradication (Banwo et al., 2017).

In 2018, thousands of online entities, both in the EU and abroad, have proactively shuttered their European operations for fear of getting caught in the GDPR's regulatory crosshairs. Meanwhile the market share of the Google, Facebook, and Amazon have increased. It is estimated that only about half of applicable firms comply with the GDPR, and some report that they may never comply because of the high costs. Official European data suggests that the regulation has not improved the operating environment for SMEs, and indeed, may not be the correct framework to resolve the reported business challenges of SMEs.

This chapter reviews those consequences and urges caution about adopting GDPR-style measures and highlights the need for careful attention in crafting any new data protection rules. It is likely that people misunderstand the GDPR, mistaking it as a way to protect privacy when instead of a method to govern data. To promote SMEs, policymakers should consider incentives, not just punishments.

3.2 Theories of Data Protection

3.2.1 Privacy vs. Data Protection

A popular misconception about the GDPR is that it protects privacy; in fact, it is about data protection or, more correctly, data governance (Evidon, 2017). The word "privacy" does not even appear in the final text of the GDPR, except in a footnote (GDPR n.18). Data privacy is about the use of data by people who are allowed to have it. Data protection, on the other hand, refers to technical systems that keep data out of the hands of people who should not have it. By its very name, the GDPR regulates the processing of personal data, not privacy.

Privacy is a complex notion having to do with being apart from others, being concealed or secluded, being free from intrusion, being let alone, and being free from publicity, scrutiny, surveillance, and unauthorized disclosure of one's personal information (See Dictionary.com, 2018). Data privacy is the application of these principles to information technology. The International Association of Privacy Professionals (IAPP) Glossary notes that data or information privacy is the "claim of individuals, groups or institutions to determine for themselves when, how and to what extent information about them is communicated to others" (IAPP, 2018a). Data protection, on the other hand, is the safeguarding of information from corruption, compromise, or loss. IPSwitch summarizes the difference: "data protection is essentially a technical issue, whereas data privacy is a legal one" (Robinson, 2018). It is important to make this distinction because the terms are often used interchangeably in popular discourse but do not, in fact, mean the same thing.

Yet some assert that the GDPR is somehow a morally superior regime, conflating the high-minded value of privacy with a secular set of technical requirements on data protection (Krishnan, 2018). The Data Protection Supervisor, the new EU super-regulator for data protection, bills itself as the "global gold standard," even though the components of the regulation that created it are relatively new and still being tested in both the marketplace and the courts (EDPS, 2018). The GDPR itself declares in Recital 4, "The processing of personal data should be designed to serve mankind (see GDPR Recital 14)." Despite EU assertions to the contrary, there are many technical forms of data protection; each has its own features, but there is no one regime, which is objectively and empirically "best."

3.2.2 Geopolitical Goals of the GDPR

The GDPR can be examined in the context of a heightened pro v. anti-EU debate, fueled by a rise in Euroscepticism[1] and nationalist parties, which

---

[1]Euroscepticism is the notion that the European integration undermines the national sovereignty of its members states, that the EU lacks democratic legitimacy, is too bureaucratic, encourages high migration, and the perception that it is a neoliberal organization benefitting the elite at the expense of the working class—remains an obstacle to the goals some have for the European continent. See also Dalibor Rohac, Europe's Pressure Points, AEI, January 17, 2017, http://www.aei.org/feature/europes-pressure-points/

charge that European integration weakens national sovereignty (FitzGibbon et al., 2016). Smarting from a disgruntled electorate and the Brexit bombshell (ibid), pro-European coalitions support pan-European regulation such as the GDPR to legitimize the EU project. It should be noted that Eurosceptic political actors are not necessarily opposed to data protection regulation; they merely prefer the primacy of national institutions over European ones, largely because of concerns that EU institutions and policies are subverting democracy.

In the case of the GDPR, there was no groundswell of public support calling for the enactment of greater data protection regulation. The GDPR was enacted during a period of voter "disengagement." (Curtice, 2016) Participation in European Parliament elections has dwindled from 62 percent in 1979 to just 42 percent in 2014 (Olson, 1971). This environment of voter disengagement is conducive for the collective action of organized special interests to defeat a diffuse, disgruntled, and unorganized majority (Olson, 1971). Relatively few Europeans are even aware of the GDPR. For example, a United Kingdom survey found that only 34 percent of respondents recognized the law, and even fewer knew what it covered (Cooke, 2018). Essentially, a relatively small group of GDPR advocates successfully implemented massive pan-European regulation without significant voter buy-in. Public opinion as measured by the Eurobarometer poll (see European commission), suggests that most people would prefer a more nuanced approach to data protection over the sledgehammer of the GDPR, and that most would rather strengthen regulation at the nation-state level than at the EU (Layton, 2017). Nevertheless, the GDPR automatically supersedes national law. If one country rules in a GDPR case in its own court, it can be overruled by a majority of EU nations.

A related geopolitical issue is the sense among Europeans that they have fallen behind U.S. and China in the internet economy (Willy, 2018). The EU continues to watch the U.S. and increasingly China, capture the world market for internet innovation and revenue. A European company has not appeared on Mary Meeker's list of top internet companies since 2013 (Kleiner Perkins, 2018).

Given that so many foreign firms have exited the EU market, it is possible that the GDPR achieved a goal in part by reducing competition. As such, the GDPR is a subset of trade policy and acts like a tariff (Lyons, 2018). European firms now have fewer US based competitors. However, this observation needs to be balanced by the fact that the market shares of the largest US firms have increased.

3.2.3 Data Protection and Cultural Norms

Different approaches to data protection and data privacy is underscored by demonstrated cultural differences and exigencies. For example, the Nordic countries, with their traditions of transparency and egalitarianism, have long maintained digital public databases of individual citizens' salary (Reuters, 2016), and income tax records (Collinson, 2018; Statiska CentralByrån, 2018) (SSB). This disclosure of financial information contradicts other national traditions and strict laws on the protection of financial information. Some countries make criminal records available to the public at the federal, state, and county level (Jacobs & Crepet, 2012), whereas such information is not available in the same way across the EU. While telephone books and White and Yellow Pages have been around for decades, had they been invented in today's precautionary environment, it is doubtful that such valuable tools would be allowed in the EU. These differences and similarities demonstrate a key debate in the field of internet policy: the individual's right to privacy versus the public's right to know (Cate et al., 1994).

Many academic studies have documented cultural differences in opinions about privacy and their implications for policy (Hainsworth, 2016). The existence of these cultural differences suggests that exporting the GDPR's one-size-fits-all approach to other nations with digital platforms may not be optimal for realizing what those other countries want in terms of data protection (Chakravorti, 2018). Consider Professor Geert Hofstede's study of cultural dimensions of citizens of the U.S. and Germany and the potential implications for data protection (Hofstede Insights, 2018). Americans score highly on individualism, geographical mobility, interacting with people they do not know, and seeking information from others. This could explain why Americans are more comfortable with sharing information, as they anticipate benefits from doing so. Germans, in contrast, score highly on uncertainty avoidance and may be more cautious with information sharing. That the leading architects of the GDPR are German and Austrian could reflect a cultural desire to lessen or avoid what they see as uncertainty in the data-driven economy, whereas Americans may believe the benefits of sharing information in society today outweigh the risks of imperfect information about the future. These conclusions regarding the different preferences for caution when disclosing data have been noted by Professors Robert Thomson (see Thomson et al., 2015) and Steven Bellman (see Bellman et al., 2004). Furthermore, studies of privacy behavior find that it is not monolithic even within cultures. Privacy concerns can diminish with education and experience

(Harris et al., 2003; Hoffman et al., 1999; Reed et al., 2016). A nation's policy choices on data privacy and protection are imbued at least to some extent with the local and culturally relevant preferences (Cockcroft, 2017).

The conflicting theoretical views regarding data privacy and protection are well summarized in Adam Thierer's *Permissionless Innovation: The Continuing Case for Comprehensive Technological Freedom (Theirer, 2016)*. He describes the precautionary principle as the belief that "innovations should be curtailed or disallowed until their developers can prove they will not cause any harm to individuals, groups, specific entities, cultural norms, or various existing laws, norms or traditions," and contrasts it with permissionless innovation, in which "experimentation with new technologies and business models should be generally permitted by default" unless a "compelling" case can be made that an innovation will bring serious harm (ibid). The EU is following the precautionary principle by enacting and enforcing the GDPR, while the U.S. subscribes to permissionless innovation by allowing innovation unless and until it has proved harmful. While the EU has deemed certain data practices presumptively harmful, it has not proved the alleged harm.

3.2.4 Data Protection and Online Trust

The GDPR could be justified if there were evidence that the many European internet-regulation laws to date have created greater trust in the digital ecosystem, but there is no such evidence. After a decade of GDPR-type regulations—in which Europeans have endured intrusive pop-ups and disclosures on every digital property they visit—Europeans report no greater sense of trust online (Castro & McQuinn, 2014). As of 2017, only 32 percent of Europeans shop outside their own country (a paltry increase of 10% in a decade), demonstrating that the European Commission's Digital Single Market (DSM) goals are still elusive (EC, 2018a). Moreover, only 20 percent of EU companies are highly digitized (EC, 2018b). These are primarily large firms. Small to medium sized companies invest little to modernize and market to other EU countries (EC, 2015). The EU has not yet offered to provide any measure that the GDPR is working to create greater trust.

Regulatory advocates would likely describe most Facebook users as suffering from a "privacy paradox" (understanding the value of privacy but failing to practice privacy enhancing behaviors) (see Wittes & Liu, 2001), but the reality may be more complex. Users interpret privacy within a context, and they do not object to sharing information per se, only to sharing that is inappropriate based on the context (Nissenbaum, 2009). Many users get value from Facebook; they like having their family and friends, photo albums, and messaging all in one place. They likely understand that advertising and data collection underpin the platform and make the valuable services possible, just as advertising supported analog television, radio, and print in the past. Naturally, users expect to be treated well, but they do not necessarily expect that platform providers will never make mistakes. Indeed, users could be upset about Cambridge Analytica, but rather than quitting Facebook, they would like to see how Facebook responds to the situation by making improvements to the platform. This may be related to Facebook having a resilient "brand personality" such that users understand that it is an imperfect and evolving platform (Aaker, 1997). Indeed, Facebook experienced an increase in engagement from U.S. users following the Cambridge Analytica revelation, as users went online to change their privacy settings (Turdiman, 2018).

However, many U.S. users do quit Facebook. Hill Holliday's survey of Generation Z (those born since 1994) shows that so-called digital natives, who are estimated to comprise 40 percent of U.S. consumers by 2020 and of whom more than 90 percent use social media platforms, found that more than one-half had switched off social media for extended periods and one-third had canceled their social media accounts (Holliday, 2018). Users cited time wasting as the reason for quitting twice as often as a concern about privacy. While service providers don't like the high rates of churn on their platforms (see Hwong, 2017), they are indicative of a competitive market in which consumers find it easy to leave and try other platforms with different features.

Additionally, reports suggest that some forms of user engagement are declining (Erskine, 2018). This could be related to Facebook changing its model to emphasize posts from family and friends over news. The most significant market response was the company losing $119 billion following its second quarter financial results, the biggest market value drop for a company on a single day in U.S. history (Imbert & Francolla, 2018). This amount is roughly 10 times the maximum fine that authorities could levy on the company under the GDPR. Moreover, Facebook's shareholders have demanded leadership changes (Trillium Asset Management, 2018) and have lodged lawsuits against the company (Bowcott & Hern, 2018). The response demonstrates that users and the marketplace can be effective regulators and is consistent with the literature about corporate response to public relations disasters such as the Tylenol scare and plane crashes: firms take steps to improve safety, frequently without being compelled by government to do so (Hazlett et al., 2018).

3.3 Preliminary Outcomes

3.3.1 Unintended Consequences

In the months since the GDPR took effect, there have been reports of European startups closing (see Kottasová, 2018) foreign news outlets pulling out of the EU (Latimes, 2018), the disruption of online ad markets (Davies, 2018), and personal inboxes being flooded with compliance emails (Hern, 2018). There are related and significant concerns about free speech, security threats, compliance costs, and innovation deterrence. Since the GDPR went into effect, over 1,000 news sites have gone dark in the EU (South, 2018).

Given the scope of Google's advertising platform and its affiliates on syndicated networks, its compliance with the GDPR has caused ripple effects in ancillary markets. Independent ad exchanges noted prices plummeting 20 to 40 percent (Davies, 2018). Some advertisers report being shut out from exchanges (Armitage, 2018). The GDPR's complex and arcane designations for "controllers" and "processors" can ensnare third party chip makers, component suppliers, and software vendors which have never interfaced with end users, as European courts have ruled that any part of the internet ecosystem can be liable for data breaches (EU, 2018).

Some GDPR requirements are fundamentally incompatible with big data, artificial intelligence, blockchain, and machine learning, especially those that require data processors to disclose the purpose of data processing, minimize their use of data, and automate decision-making (Zarsky, 2017). For technology developers, engineers, and entrepreneurs, the GDPR creates uncertainty not only in the text of the law and its adjudication, but in that requirements and tenets of the GDPR conflict with the operation of machine learning and artificial intelligence (Thayer & Madhani, 2018).

Some of the most important recent scientific advances have been the result of processing various sets of information in inventive ways—ways that neither subjects nor controllers anticipated, let alone requested. Consider the definitive study on whether the use of mobile phones causes brain cancer (Frei et al., 2011). The Danish Cancer Society analyzed 358,403 Danish mobile subscribers by processing social security numbers, mobile phone numbers, and the National Cancer Registry, which records every incidence of cancer by social security number (ibid). The study, the most comprehensive investigation of its kind ever conducted, found no correlation of the use of mobile phones with brain cancer. But the users' information was not collected for the express purpose of such a study. Therefore, it's possible that, had the GDPR been in effect at the time of the study, consent from the population whose data was analyzed would not have been available, and the GDPR's purpose-specification requirement would have therefore made it impossible to conduct the study. Going forward, it's possible, if not likely, that valuable research will not be conducted because of the GDPR.

Indeed, part of the promise of socialized medicine was the ability to tap the vast pools of data in public health databases to make advances in medicine. However, a privacy panic is threatening to derail some projects (see Castro & McQuinn, 2015) including Iceland's genome warehouse, the oldest and most complete genetic record in the world, which promises groundbreaking therapies for Alzheimer's disease and breast cancer (Hsu, 2015). While many regulatory advocates focus attention on Silicon Valley firms and call for greater regulation, their campaign is backfiring as users turn their ire toward governments and demand erasure of their data from national health care records and other government services, potentially frustrating the operating models of mandated social programs (Howell, 2018). With the mantra of "if in doubt, opt out," about half a million Australians rejected the country's national electronic health record, causing the computer system to crash in July 2018 (ibid).

For centuries, European state churches have collected and published information on births, deaths, weddings, baptisms, and more. In Denmark and Sweden, these institutions retain the official register for this information. Because of the GDPR, many churches have stopped printing announcements in the bulletins for their local congregations unless they obtain consent first (Version2, 2018)(BT, 2018)(østergaard, 2018). GDPR risks have also been identified with respect to convicted felons successfully removing information about their crimes from search engines[2] (Castro, 2018), the exchange of business cards (White, 2018), the taking of pictures in public (Nelson, 2018; Sullivan, 2018), and disclosures of health and injury information in the trade of soccer players (Idskov, 2018). In a remarkable example of "evasive innovation" in which entrepreneurs find workarounds to law and regulation as form of civil disobedience (Thierer, 2018), the browser company Opera,

---

[2]"According to the company (Google), almost one-fifth of the news articles it received requests to remove related to crime, and it removes roughly one-third of the right to be forgotten requests that it receives relating to news articles."

at user request, has developed a download blocker for to stop the mandated GDPR disclosures which consume data and battery life on mobile phones (Popa, 2018).

A key unintended consequence of the GDPR is that it undermines the transparency of the international systems and architecture that organize the internet. The WHOIS query and response protocol for internet domain names, IP addresses, and autonomous systems is used by law enforcement, cybersecurity professionals and researchers, and trademark and intellectual property rights holders (Tews, 2018a). The Internet Corporation for Assigned Names and Numbers (ICANN) recently announced a Temporary Specification that allows registries and registrars to obscure WHOIS information they were previously required to make public, ostensibly in order to comply with the GDPR (ICANN, 2018a). This could hinder efforts to combat unlawful activity online, including identity theft, cyber-attacks, online espionage, theft of intellectual property, fraud, unlawful sale of drugs, human trafficking, and other criminal behavior, and it is not even required by the GDPR.

The GDPR does not apply at all to non-personal information and states that disclosure of even personal information can be warranted for matters such as consumer protection, public safety, law enforcement, enforcement of rights, cybersecurity, and combating fraud. Moreover, the GDPR does not apply to domain names registered to U.S. registrants by American registrars and registries. Nor does it apply to domain name registrants that are companies, businesses, or other legal entities, rather than "natural persons." All the same, actors including ICANN are practicing voluntary censorship because the GDPR's provisions are so vague and the potential penalties so high. GDPR proponents have likely contributed to the impression that the GDPR urges measures like the Temporary Specification. For example, in her role in the Article 29 Working Party, the group that drove the promulgation of the GDPR, Jelinek said that the elimination and masking of WHOIS information is justified under the GDPR (ICANN, 2018b).

The WHOIS problem can be described as the conflict between the individual's right to privacy and the public's right to know (Tews, 2018b). It can also be understood within the context of the problem of "privacy overreach" (see Hurwitz and Jaffer, 2018), in which the drive to protect privacy becomes absolute, lacks balance with other rights, and unwittingly brings worse outcomes for privacy and data protection (Brkan, 2016). The situation harkens back to a key fallacy of privacy activists who attempted to block the rollout of caller ID because it violated the privacy rights of intrusive callers. Today, the receiver's right to know who is calling is prioritized over the caller's right to remain anonymous (Hurwitz & Jaffer, 2018). Similarly it is understood that the needs of public safety will supersede data protection, particularly in situations of danger to human life. Moreover, one should expect intellectual property to be in balance with data protection, not in conflict as it is under the GDPR. The pace of development of privacy and data protection law is significantly faster than that of other kinds of law, leading one scholar to suggest that it threatens to upend the balance with other fundamental rights (Brkan, 2016). Moreover the expansion of rights comes with social consequences of the imposition of correlative on others (Epstein, 2018).

3.3.2 Impacts to SMEs

One disturbing outcome from the regulation is the negative impact on venture capital for SMEs in the EU. In contrast to GDPR policymakers' pronounce-ment that the regulation would be good for SMEs, it appears to be the opposite. A recent study compared venture capital investments within SMEs in the respective EU and US data collection spaces from July 2017–Sept 2018 (Jia et al., 2018). It found in the EU a $3.38M decrease in total dollars raised per country per week; a 17.6 percent reduction in weekly venture deals; and a 39.6 percent decrease in the amount raised per deal. These declines are estimated to reduce 3000–30,000 jobs lost. A study by a German digital marketing research firm noting that Google is the biggest beneficiary of the GDPR, reports that European ad tech firms have fallen in rank and market share by 18–32 percent (Greif, 2018). Indeed various entities report that the decline in SMEs and the retraction from the market has increased the respective shares of Google, Facebook and Amazon (Scott et al., 2018).

These results are consistent with annual reports from European authori-ties. The annual Digital Scoreboard reports from the European Commission note the moribund performance of European SMEs. GDPR type regulation have been in place for at least a decade in parts of the EU, but the growth in the percent of SMEs that sell their products and services online been "glacial", from 13% in 2010 to 18% of SMEs in 2017 and little to no improvement has been observed in their revenue or their likelihood that they sell outside their own country (EC 2018c). In 2017 ecommerce comprised just 13 percent of revenue for medium firms of 50–249 and just 7.4 percent for firms with less than 49 employees. There are differences across countries of course as Ireland and Czech Republic perform better, but most countries perform at or below the average with Romania and Bulgaria being the laggards. The situation may also reflect the low level of integration of information technologies in general, as SMEs particularly in Romania, Bulgaria, Hungary, Latvia, Italy, and Poland may have just a basic website and a computer or two.

Big data, particularly open source technologies, has been celebrated as a means for innovation by small players (Vecchio et al., 2018), but the use of big data also diverges between SMEs and large enterprise, with only 10 percent of SMEs taking advantage of the technology while large firms are more than twice as likely to do so.

These data suggest the difficulties imposed by the GDPR, for if SMEs have a difficult to procure basic IT equipment, it is even more unlikely that the can implement sophisticated data protection practices. Moreover, the data shows a correlation between businesses that use social media and other interactive technologies to engage with customers and a propensity to digitize the enterprise, a dynamic used by large enterprises already in the EU. GDPR likely increases the business risks for SMEs to employ such practices and hence they are discouraged from some kinds of digitization.

3.3.3 Compliance Costs

Prior to its rollout, GDPR was expected to cost firms with 500 employees or more will likely have to spend between USD$1 million and $10 million each (PricewaterhouseCoopers, 2017). To date, the costs total $3 million each for Fortune 500 firms in the US and UK (Smith, 2018). The initial average outlay was $1.3 million with another $1.8 million expected to be fully compliant. Many of these costs are being realized as increased fees from IT vendors (Carson, 2018). The IAPP-EY Annual Privacy Governance Report 2018 notes that less than 50% of firms report being fully compliant with the GDPR and 20 percent say full compliance is impossible (IAPP, 2018b).

Another study estimates the total cost for all relevant American businesses at $150 billion (U.S. Census Bureau, 2015), twice what the U.S. spends on network investment (Spalter, 2017) and one-third of annual e-commerce revenue in the U.S. (U.S. Census Bureau, 2018). Economist Hosuk Lee-Makiyama calculates that the GDPR's requirements on cross-border trade flows will increase prices, amounting to a direct welfare loss of €260 per European citizen[3] (Lee-Makiyama, 2014). The net effect is that those companies that can afford to comply will do so, and the rest will exit. Hence the GDPR will become a barrier to market entry, punishing small firms, rewarding the largest players, and creating a codependent relationship

---

[3]This methodology is expanded in Erik Van der Marel et al., A Methodology to Estimate the Costs of Data Regulations, 146 INT'L ECON. 12 (2016).

between regulators and the firms they regulate. This is a perverse outcome for a regulation that promised to level the playing field on data protection.

The GDPR imposes massive new responsibilities on regulators without a concurrent increase in training, funding, and other resources. EU data supervisors wear many hats, including "ombudsman, auditor, consultant, educator, policy adviser, negotiator, and enforcer" (Bennet & Raab, 2006). Furthermore, the GDPR widens the gap between the high expectations for data protection and the low level of skills possessed by data supervisors charged with its implementation (Raab & Szekely, 2017). There are certainly many talented individuals among these ranks, but the mastery of information and communications technology varies considerably among these professionals, especially as each nation's DPA is constituted differently.

3.4 Conclusion

This chapter has reviewed some theories of data protection as they related to Europe's GDPR and preliminary outcomes for SMEs. Developing policymakers should take a critical look at GDPR outcomes particularly in policy development for SMEs. The scientific research on data protection and privacy suggests that consumer education and privacy enhancing technologies are essential to creating trust online (Layton, 2018) but these inputs were ignored in developing the GDPR (Cal. Civ. Code §1798.100)[4].

The purpose of the GDPR is not to protect privacy, but rather to regulate data processing. In the past decade, the increasing data protection rules have not resulted in improved trust, increased cross-border commerce in the EU, or the growth of SMEs in the digital sector. Compliance costs are so high that many foreign firms have stopped serving the EU, some have closed all together, and many firms decide not to use digital technologies.

Notably US policymakers have noted some of the detrimental effects of the policy and therefore attempt to develop an innovation-based data protection framework which will not harm SME innovators. Presently the US is engaged in a process of public comment and scientific inquiry to develop a set of principles that will provide a high level of protection for individuals while giving organizations legal clarity and the flexibility to

---

[4]Many compare this new state law to the GDPR because of its heavyhanded approach and potentially negative impact for enterprise. See Lothar Determan, Analysis: California Consumer Privacy Act of 2018, IAPP, July 2, 2018, https://iapp.org/news/a/analysis-the-california-consumer-privacy-act-of-2018/.

innovate (NTIA, 2018). Developing country policymakers should consider balanced policies that promote both privacy and prosperity. Rather than copy the GDPR, policymakers should consider how consumers can access privacy education to make informed choices and how to implement safe harbors for privacy-enhancing innovation and protecting the testing and learning of new technologies (ibid).

References

Aaker, J. L. (1997). Dimensions of brand personality, *J. Marketing Res.,*34(3), 347–356. http://www.haas.berkeley.edu/groups/finance/Papers/Dimensions%20of%20BP%20JMR%201997.pdf

Armitage, C. (2018). Life after GDPR: what next for the advertising industry?, World Federation Of Advertisers, https://www.wfanet.org/news-centre/life-after-gdpr-what-next-for-the-advertising-industry/.

Ashwin, K. (2018). GDPR Is Not Just a Regulatory Framework. It's Also a Moral and Existential Blueprint, CSO ONLINE, https://www.csoonline.com/article/3257695/privacy/gdpr-is-not-just-a-regulatory-framework-it-s-also-a-moral-and-existential-blueprint.html.

Banwo, A. O., Du, J. and Onokala, U. J. (2017). The determinants of location specific choice: small and medium-sized enterprises in developing countries. *Glob Entrepr Res.*, 7(1), 16.

Bellman, S., Johnson, E. J., Kobrin, S. J. and Lohse, G. L. (2004). International Differences in Information Privacy Concerns: A Global Survey of Consumers, *The Info. Soc'y.*, 20(5), 313–324.

Bennett, C. J., and Raab, C. (2006). *The Governance of Privacy: Policy Instruments in Global Perspective*. The MIT Press.

Bowcott, O. and Hern, A. (2018). Facebook and Cambridge Analytica Face Class Action Lawsuit, THE GUARDIAN, https://www.theguardian.com/news/2018/apr/10/cambridge-analytica-and-facebook-face-class-action-lawsuit.

Brkan, M. (2016). The Unstoppable Expansion of the EU Fundamental Right to Data Protection, 23 *Maastricht J. Of Euro. & Comp. Law.*, 812, http://journals.sagepub.com/doi/abs/10.1177/1023263X1602300505?journalCode=maaa.

BT. (2018). Kirkeblade opgiver at bringe navne pådøbte og døde, B.T., https://www.bt.dk/content/item/1203799;

Cal. Civ. Code §1798.100 et seq.

Cate, F., Fields, D. and McBain, J. (1994). The Right to Privacy and the Public's Right to Know: The 'Central Purpose' of the Freedom of Information Act, *Admin. L. Rev,* 46, 41. https://www.repository.law.indiana.edu/facpub/737.

Carson, A. (2018). "Should vendors be able to pass along costs of GDPR compliance?" IAPP. https://iapp.org/news/a/should-vendors-be-able-to-pass-along-costs-of-gdpr-compliance/

Castro, D. (2018). The EU's Right to be Forgotten is Now Being Used to Protect Murderers, CENTER FOR DATA INNOVATION, https://www.datainnovation.org/2018/09/the-eus-right-to-be-forgotten-is-now-being-used-to-protect-murderers).

Castro, D. and McQuinn, A. (2014). The Economic Cost of the European Union's Cookie Notification Policy, *ITIF*, https://itif.org/publications/2014/11/06/economic-cost-european-unions-cookie-notification-policy.

Castro, D. and McQuinn, A. (2015). The Privacy Panic Cycle: A Guide to Public Fears about New Technologies. Information Technology and Innovation Foundation, https://itif.org/events/2015/09/10/sky-not-falling-understanding-privacy-panic-cycle.

Census Bureau. (2018). 2015 SUSB Annual Data Tables by Establishment Industry, https://www.census.gov/data/tables/2015/econ/susb/2015-susb-annual.html.

Chakravorti, B. (2018). Why the rest of the world can't free ride on the GDPR, *Harv. Bus. Rev*., https://hbr.org/2018/04/why-the-rest-of-world-cant-free-ride-on-europes-gdpr-rules.

Cockcroft, S. (2007). Culture, Law and Information Privacy, *Proceedings of European and Mediterranean Conference on Information Systems, Polytechnic University of Valencia, Spain*, http://emcis.eu/Emcis_archive/EMCIS/EMCIS2007/emcis07cd/EMCIS07-PDFs/642.pdf.

Collinson, P. (2016). Norway, the Country Where You Can See Everyone's Tax Returns, THE GUARDIAN, https://www.theguardian.com/money/blog/2016/apr/11/when-it-comes-to-tax-transparency-norway-leads-the-field;

Cooke, K. (2018). *Data Shows Awareness of GDPR Is Low amongst Consumers*. KANTAR, https://uk.kantar.com/public-opinion/policy/2018/data-shows-awareness-of-gdpr-is-low-amongst-consumers/.

Curtice, J. (2016). How Deeply Does Britain's Euroscepticism Run? NATCEN, http://www.bsa.natcen.ac.uk/media/39024/euroscepticism.pdf.

David, R. (2018). Data Privacy *vs*. Data Protection, IPSWITCH, https://blog.ipswitch.com/data-privacy-vs-data-protection.

Davies, J. (2018). 'The Google Data Protection Regulation': GDPR is Strafing Ad Sellers, DIGIDAY, https://digiday.com/media/google-data-protection-regulation-gdpr-strafing-ad-sellers/.

EC. Public Opinion. European Commission, http://ec.europa.eu/commfront office/publicopinion/index.cfm.

EDPS. *The History of the General Data Protection Regulation.* European Data Protection Supervisor, https://edps.europa.eu/data-protection/data-protection/legislation/history-general-data-protection-regulation_en (acces sed September 27, 2018).

Epstein, R. A. (2018). A Not Quite Contemporary View of Privacy, *Harv. J. Pub. Pol.*, 41, 95, http://www.harvard-jlpp.com/wp-content/uploads/2018/01/EpsteinPanel_FINAL.pdf.

Erskine, R. (2018). Facebook Engagement Plunging Over Last 18 Months, FORBES, https://www.forbes.com/sites/ryanerskine/2018/08/13/study-facebook-engagement-sharply-drops-50-over-last-18-months/#69c1de579 4e8.

EU. (2018). *Judgment of the Court (Grand Chamber), EU,* https://eur-lex.europa.eu/legal-content/EN/TXT/HTML/?uri=CELEX:62016CJ0210& qid=1531145885864&from=EN.

European Commission. (2015). Better Access for Consumers and Business to Online Goods, https://ec.europa.eu/digital-single-market/en/better-access-consumers-and-business-online-goods.

EVIDON. (2017). *What Is the GDPR?* (last visited Aug. 25, 2017), https://www.evidon.com/education-portal/videos/what-is-the-gdpr

European Commission. (2018). *2018 Reform of EU Data Protection Rules.* https://ec.europa.eu/commission/priorities/justice-and-fundamental-rights/data-protection/2018-reform-eu-data-protection-rules_en.

European Commission. (2018a). Use of Internet Services. http://ec.europa. eu/information_society/newsroom/image/document/2018-20/3_desi_report _use_of_internet_services_18E82700-A071-AF2B-16420BCE813AF9F0_5 2241.pdf. See id. at 4 ("Growth in the use of online services is generally slow.").

European Commission. (2018b). Integration of Digital Technology. http://ec.europa.eu/information_society/newsroom/image/document/2018-20/4_desi_report_integration_of_digital_technology_B61BEB6B-F21D-9D D7-72F1FAA836E36515_52243.pdf.

European Commission. (2018c). "Digital Economy and Society Index Report 2018 – Integration of Digital Technology". https://ec.europa.eu/digital-single-market/en/integration-digital-technology

FitzGibbon, J., Leruth, B. and Startin, N. (2016). *Euroscepticism as a Transnational and Pan-European Phenomenon*. Routledge, Taylor and Francis group.

Frei, P, Poulsen, A. H., Johansen, C., Olsen, J. H., Steding-Jessen, M., and Schüz, J. (2011). Use of Mobile Phones and Risk of Brain Tumours: Update of Danish Cohort Study, *BMJ,* https://www.cancer.dk/dyn/resou rces/File/file/9/1859/1385432841/1_bmj_2011_pdf.pdf.

GDPR, n.18 (referring to Directive 2002/58/EC of the European Parliament and of the Council of 12 July 2002 concerning the processing of personal data and the protection of privacy in the electronic communications sector).

GDPR, Recital 4.

Greif, B. (2018). Google is the biggest beneficiary of the GDPR. Cliqz. https://cliqz.com/en/magazine/study-google-is-the-biggest-beneficiary-of-the-gdpr

Hainsworth, J. (2016). *Global Privacy Ethics Subject to Cultural Differences*, BNA, https://www.bna.com/global-privacy-ethics-n57982069807/.

Harris, M. M., Van Hoye, G. and Lievens, F. (2003). Privacy and Attitudes Towards Internet-Based Selection Systems: A Cross-Cultural Comparison, *Int'l J. Selection & Assessment,* 11, 230.

Hazlett, T., Jaffer, J., Stifel, M. and Heiman, M. (2018). What to do about Facebook: On Data Privacy and the Future of Tech Regulation, REGULATORY TRANSPARENCY PROJECT TELEFORUM POD-CAST, https://fedsoc.org/events/RTP_FTC-FB-CamAnalytica (comments of economist Thomas Hazlett).

Hern, A. (2018). Most GDPR Emails Unnecessary and Some Illegal, Say Experts, THE GUARDIAN (May 21, 2018). https://www.theguardian.com/technology/2018/may/21/gdpr-emails-mostly-unnecessary-and-in-some-cases-illegal-say-experts.

Hoffman, D. L., Novak, T. P. and Peralta, M. A. (1999). Information Privacy in the Marketspace: Implications for the Commercial Uses of Anonymity on the Web, *The Info. Soc'y.,* 15, 129.

Hofstede Insights. Country Comparison, https://www.hofstede-insights.com/country-comparison/ (accessed September 27, 2018).

Holliday, H. (2018). Meet Gen Z: The Social Generation, http://thinking.hhcc.com/?utm_campaign=Thought%20Leadership%20%E2%80%94%20Gen%20Z&utm_source=

Howell, B. (2018). Data Privacy Debacle Down Under: Is Australia's My Health Record Doomed? AEIdeas, http://www.aei.org/publication/data-privacy-debacle-down-under-is-australias-my-health-record-doomed/

Hsu, J. (2015). Iceland's Giant Genome Project Points to Future of Medicine, *IEEE SPECTRUM*, https://spectrum.ieee.org/the-humanos/biomedical/diagnostics/icelands-giant-genome-project-points-to-future-of-medicine.

Hurwitz, J. G. and Jaffer, J. N. (2018). Modern Privacy Advocacy: An Approach at War with Privacy Itself?, *Regulatory Transparency Project White Paper*, https://regproject.org/paper/modern-privacy-advocacy-appro ach-war-privacy/.

Hwong, C. (2017). Why Churn Rate Matters: Which Social Media Platforms Are Losing Users?, VERTO ANALYTICS, https://www.vertoanalytics. com/chart-week-social-media-networks-churn/.

IAPP. (2018a). Information Privacy, Glossary, International Association of Privacy Professionals https://iapp.org/resources/glossary/#information-privacy (accessed September 27, 2018).

IAPP. (2018b). International Association of Privacy Professionals. IAPP-EY Annual Governance Report. https://iapp.org/resources/article/iapp-ey-annual-governance-report-2018/

ICANN. (2018a). Temporary Specification for gTLD Registration Data, ICANN, https://www.icann.org/resources/pages/gtld-registration-data-specs-en

ICANN. (2018b). Letter from Andrea Jelinek, Chairperson of Article 29 Data Protection Working Party, to Göran Marby, President of ICANN, https://www.icann.org/en/system/files/correspondence/jelinek-to-marby-11apr18-en.pdf.

Idskov, T. (2018). Mundkurv! Derfor holdes omfanget af FCK-spillers skade hemmelig, B.T., https://www.bt.dk/content/item/1197424.

Imbert, F. and Francolla, G. (2018). Facebook's $100 Billion-plus Rout Is the Biggest Loss in Stock Market History, CNBC, https://www.cnbc.com/ 2018/07/26/facebook-on-pace-for-biggest-one-day-loss-in-value-for-any-company-sin.htm

Jacobs, J. and Crepet, T. (2007). The Expanding Scope, Use, and Availability of Criminal Records, *NYUJ Legis. & Pub. Pol'y.*, 11, 177, http://www.nyujlpp.org/wp-content/uploads/2012/10/Jacobs-Crepet-The-Expanding-Scope-Use-and-Availability-of-Criminal-Records.pdf.

Jia, J., Jin, G. Z., and Wagman, L. (2018). The Short-Run Effects of GDPR on Technology Venture Investment. https://www.nber.org/papers/w25248

Kleiner, P. (2018). Internet Trends Report 2018, https://www.kleinerperkins. com/perspectives/internet-trends-report-2018.

Kottasová, I. (2018). These Companies Are Getting Killed by GDPR, CNN, http://money.cnn.com/2018/05/11/technology/gdpr-tech-companies-losers/index.html

Layton, R. (2017). How the GDPR Compares to Best Practices for Privacy, Accountability and Trust, SSRN Scholarly Paper, https://papers.ssrn.com/abstract=2944358.

Layton, R. (2018). *Statement before the Federal Trade Commission on Competition and Consumer Protection in the 21st Century Hearings,* Project Number P181201, Market Solutions for Online Privacy, AEI, https://www.ftc.gov/system/files/documents/public_comments/2018/08/ftc-2018-0051-d-0021-152000

Lee-Makiyama, H. (2014). The Political Economy of Data: EU Privacy Regulation and the International Redistribution of Its Costs, in *Protection Of Information And The Right To Privacy: A New Equilibrium,* 85–94.

Los Angeles Times, TRONC (last visited June 25, 2018), http://www.tronc.com/gdpr/latimes.com/.

Lyons, D. (2018). GDPR: Privacy as Europe's tariff by other means?. *American Enterprise Institute.* https://www.aei.org/publication/gdpr-privacy-as-europes-tariff-by-other-means/

Nissenbaum, H. (2009). *Privacy In Context: Technology, Policy, And the Integrity of Social Life.* Standford University Press, https://www.sup.org/books/title/?id=8862.

NTIA. (2018). Request for Comments on Developing the Administration's Approach to Consumer Privacy, NATIONAL TELECOMMUNICATIONS AND INFORMATION ADMINISTRATION, https://www.ntia.doc.gov/federal-register-notice/2018/request-comments-developing-administration-s-approach-consumer-privacy.

Nelson, S. S. (2018). New EU Data Protection Law Could Affect People Who Take Pictures with Their Phones, *NPR*, https://www.npr.org/2018/05/24/614195844/new-eu-data-protection-law-could-affect-people-who-take-pictures-with-their-phon?t=1538121870256.

Olson, M. (1971). *The Logic of Collective Action.* Harvard University Press.

Østergaard, J. P. (2018). Konsekvens af EU-lov: Slut med at læse om døbte, gifte og døde, VIBORG STIFTS FOLKEBLADE, https://viborg-folkeblad.dk/rundtomviborg/Konsekvens-af-EU-lov-Slut-med-at-laese-om-doebte-gifte-og-doede/artikel/376140.

Popa, B. (2018). Opera for Android Can Now Block Those Annoying GDPR Dialogs. Softpedia. https://news.softpedia.com/news/opera-for-android-can-now-block-those-annoying-gdpr-dialogs-523636.shtmlPress%20ReleaSe.

PricewaterhouseCoopers, (2017). GDPR Compliance Top Data Protection Priority for 92% of US Organizations in 2017, According to PwC Survey, https://www.pwc.com/us/en/press-releases/2017/pwc-gdpr-compliance-press-release.html.

Privacy, dictionary.com, https://www.dictionary.com/browse/privacy (accessed September 27, 2018).

Raab, C. D. and Szekely, I. (2017). Data Protection Authorities and Information Technology, *Computer L. & Sec. Rev.*, https://ssrn.com/abstract=2994898.

Reed, P. J., Spiro, E. S. and Butts, C. T. (2016). Thumbs up for privacy?: Differences in online self-disclosure behavior across national cultures, *Social Sci. Research,* 155.

Regulation (EU) 2016/679 of the European Parliament and of the Council of 27 April 2016 on the Protection of Natural Persons with Regard to the Processing of Personal Data and on the Free Movement of Such Data, and Repealing Directive 95/46/EC (General Data Protection Regulation) (Text with EEA Relevance), Pub. L. No. 32016R0679, 119 OJ L, Recital 1, Article 1 (2016), http://data.europa.eu/eli/reg/2016/679/oj/eng (hereinafter GDPR).

Reuters. (2016). Privacy, What Privacy? Many Nordic Tax Records Are a Phone Call Away, REUTERS, https://www.reuters.com/article/us-panama-tax-nordics-idUSKCN0X91QE.

Scott, M., Cerulus, L. and Kayali, L. (2018). Six months in, Europe's privacy revolution favors Google, Facebook. Politico. https://www.politico.eu/article/gdpr-facebook-google-privacy-data-6-months-in-europes-privacy-revolution-favors-google-facebook/

Smith, O. (2018). "The GDPR Racket: Who's Making Money from This $9bn Business Shakedown." Forbes. https://www.forbes.com/sites/oliversmith/2018/05/02/the-gdpr-racket-whos-making-money-from-this-9bn-business-shakedown/#1d29473134a2

South, J. (2018). More than 1,000 U.S. News Sites Are Still Unavailable in Europe, Two Months after GDPR Took Effect, NIEMAN LAB, http://www.niemanlab.org/2018/08/more-than-1000-u-s-news-sites-are-still-unavailable-in-europe-two-months-after-gdpr-took-effect/.

Spalter, J. (2018). Broadband CapEx Investment Looking Up in 2017, UST-ELECOM, https://www.ustelecom.org/blog/broadband-capex-investment-looking-2017.

SSB. (2018). Tax Statistics for Personal Tax Payers, ssb.no, https://www.ssb.no/en/inntekt-og-forbruk/statistikker/selvangivelse/aar-forelopige/2018-04-18;

STATISTISKA CENTRALBYRÅN. (2018). Income and Tax Statistics in Sweden, http://www.scb.se/en/finding-statistics/statistics-by-subject-area/household-finances/income-and-income-distribution/income-and-tax-statistics/.

Sullivan, K. (2018). What Photographers Need to Know About GDPR, PDNPULSE, https://pdnpulse.pdnonline.com/2018/06/gdpr-how-bad-is-it-for-photographers.html;

Tews, S. (2018a), How European data protection law is upending the Domain Name System, AEI, https://www.aei.org/publication/how-european-data-protection-law-is-upending-the-domain-name-system/.

Tews, S. (2018b). Privacy and Europe's data protection law: Problems and implications for the US, AEI, http://www.aei.org/publication/privacy-and-europes-data-protection-law-problems-and-implications-for-the-us/.

Thayer, J. and Madhani, B. (2018) . Can a Machine Learn Under the GDPR*?, TPRC 46: The 46th Research Conference on Communication, Information and Internet Policy,* https://ssrn.com/abstract=3141854.

Theirer, A. (2016). Permissionless Innovation: The Continuing Case For Comprehensive Technological Freedom Mercactus Center, George Mason University. Available at https://www.mercatus.org/publication/permissionless-innovation-continuing-case-comprehensive-technological-freedom.

Thierer, A. (2018). "Evasive Entrepreneurialism and Technological Civil Disobedience: Basic Definitions." The Bridge. July 20 2018. https://www.mercatus.org/bridge/commentary/evasive-entrepreneurialism-and-technological-civil-disobedience-basic-definitions

Thomson, R., Yuki, M. and Ito, N. (2015). A socio-ecological approach to national differences in online privacy concern: The role of relational mobility and trust, *Computers in Human Behavior,* 285.

Trillium Asset Management. (2019). Facebook, Inc. — Independent Board Chairman. http://www.trilliuminvest.com/shareholder-proposal/facebook-inc-independent-board-chairman-2019/ (accessed August 20, 2018).

Turdiman, D. (2018). Facebook Engagement Surge Post-Cambridge Analytica, FAST COMPANY, https://www.fastcompany.com/40563518/why-facebooks-engagement-surged-after-cambridge-analytica.

Turnout. (2014). European Parliament, European Parliament, http://www. europarl.europa.eu/elections2014-results/en/turnout.html (accessed July 27, 2018).

U.S. Census Bureau. (2018). Quarterly Retail E-Commerce Sales 1st Quarter 2018, https://www.census.gov/retail/mrts/www/data/pdf/ec_current.pdf.

Vecchio, P. D., Minin, A. D., Petruzzelli, A. M., Panniello, U. and Pirri, S. (2018). Big data for open innovation in SMEs and large corporations: Trends, opportunities, and challenges. *Creat Innov Manag.* 27, 6–22. https://onlinelibrary.wiley.com/doi/pdf/10.1111/caim.12224

VERSION2. (2018). Minister: Krav om GDPR-samtykke til kirkeblade er absurd, VERSION2, https://www.version2.dk/artikel/minister-krav-gdpr-samtykke-kirkeblade-absurd-1086182;

Wittes, B. and Liu, J. (2001). *The Privacy Paradox: The Privacy Benefits of Privacy Threats*, BROOKINGS, https://www.brookings.edu/research/the-privacy-paradox-the-privacy-benefits-of-privacy-threats/.

White, S. (2018). How Do Business Cards Sit with GDPR?, GDPR.REPORT, https://gdpr.report/news/2018/02/08/business-cards-sit-gdpr/.

Will, C. (2018). Europe's tech race – trying to keep pace with US and China, *EU OBSERVER*, https://euobserver.com/opinion/142056.

Zarsky, T. Z. (2017). Incompatible: The GDPR in the Age of Big Data, 47 SETON HALL L. REV., https://scholarship.shu.edu/cgi/viewcontent.cgi?article=1606&context=shlr.

4

Utility Cooperatives as Rural NGT Providers: Feasibility, Potentials and Pitfalls

Darío M. Goussal

School of Engineering, Northeastern University at Resistencia, Argentina
dgoussal@yahoo.com

The conventional wisdom about utility cooperatives entering in the broadband business is that of a niche operator, and a response to a market failure: since NGTs are considered attractive just by incumbent large providers operating in dense urban areas, in the absence of competitive challenges from other infrastructure providers -like utilities-, they may adopt a delayed investment policy out of those premium spots. However, the nature of utility coopera-tives is more complex to disentangle and to comprehensively encompass by scientific research. The aim of this chapter is to shed some light to feasibility aspects of utility cooperatives as potential NGT providers by examining and comparing real experiences in two countries with large rural areas and a sound tradition of utility cooperatives operating electricity and telecommunications networks: Argentina and USA, to a extent in opposite sides of development but sharing a quite similar pattern in the history and the dissemination of non-profit utilities nationwide.

4.1 Introduction

In 2014, during the 6th World Telecommunications Development Conference (WTDC-14), one of the Regional Initiatives for the Americas approved by the ITU was "Development of broadband access and adoption of broad-band", aimed at providing assistance to member states in the development of policies to increase access and uptake, wherein the expected results include support to *non-profit cooperatives* that provide services in underserved rural and suburban areas. Utility cooperatives have a long-standing tradition as reliable operators of electricity and other networked services in rural areas and remote locations where investment is neglected by investor-owned firms, in both developing countries and the industrialized world. Along history such tradition with deep roots in localism and self-management practices, typical in small and rural communities forged and consolidated loyal client bases in a range of ventures from electricity distribution to water, gas or telephone in the XX Century, and then seamlessly adding broadband connectivity in the XXI one (Settles, 2014; Sadowski, 2014; Duvall, 2016).

Moreover, due to the inherent economies of scope resulting in lower infrastructure, marketing and maintenance costs, utility cooperatives have been regarded as providers of last resort for broadband services and Next Generation Technologies in unserved or underserved areas. These providers represent a sort of "hybrid" players, different from public operators (i.e. municipal networks) and private companies (telcos, cablecos). The organizational structure, expansion policies and business behavior are different too, since cooperatives tend to operate with near or under-zero-margin profits, conservative service rates and long payback periods. Therefore, a key aspect in the study of rural utility cooperatives as broadband providers is the way they could bring NGT services while preserving overall feasibility (Mamouni et al., 2016).

In USA, there are some 900 cooperatives providing telecommunications in more than 40% of its territory, 73% of them delivering video and 66%, wireless services. Argentina has some 600 electricity and 300 telephone cooperatives from which around 250 utilities have been granted licenses to provide connectivity and broadband services, thereby with actual or expected investment plans in NGT services. However, facts and figures are opaque and difficult to get from small rural operators, and due to scattered patterns, varying sizes and services portfolios their systematization at the micro level beyond the case studies entails a lengthy, challenging task.

A research started in 2015 aims at identifying and evaluating determining factors of the feasibility of incorporation of rural broadband services in designated service areas of utility cooperatives in Argentina. It assumes the existence of utility cooperatives with infrastructure of networked public services (electricity, water, telephony, etc.) already deployed in the designated service areas, and legally entitled to add the provision of rural broadband access to their actual portfolio of services.

They are classified in four contingent situations or groups:

- Cooperatives currently providing any kind of rural broadband access (in designated service areas out of towns/urban areas of municipalities).
- Cooperatives currently providing just urban broadband accesses (yet without rural customers in designated service areas).
- Cooperatives with rural clients of other networked services in their designated service areas (electricity, telephony, etc), not yet providing urban/rural broadband.
- Cooperatives not included in groups A-B-C, willing to undertake the provision of rural broadband access.

Methodologically, this entails a techno-economic baseline study at the national level with data of sample of utility cooperatives compiled in an ad-hoc database. Such repository had been built since 2004 and currently telecom providers and electricity providers in Argentina and USA. In the rest of this chapter, selected results of the study will be discussed in a framework where the feasibility of provision of rural broadband services is analyzed according three layers or dimensions: intrinsic, institutional and customer (demand side). A previous background contains data and references to ad-hoc literature as well as in the discussion of the third layer, while the technical end engineering aspects are presented after the discussion of the first one (intrinsic feasibility). The conclusion has a few reflections on the findings of the study and a comparison outlook between the observed expansion behaviour of rural broadband projects in utility cooperatives of Argentina and USA.

4.2 Background

Occasionally each many years, scientific research provides the opportunity to make observations on strategic aspects of the telecommunications industry and to register data of selected phenomena revealing a particular behaviour, otherwise difficult to keep track of. Such was the case of the wave initiated around 2004/2005, where large groups of utility cooperatives in USA and Latin America started a silent but persistent expansion towards the provision of NGT and rural broadband services.

After more than a decade of pioneering trials and pilot projects of generally, conservative budgets and limited scope, the availability of public funding programs in USA led to the consolidation of a bundle of successful experiences and a steady rise in the number of providers there are 897 electric cooperatives serving a median of 12.000 customers, covering a vast geographical area with around 42% of the nation's distribution lines. In 2016 there were just 33 electric coops with an ongoing broadband project. In 2018, the count reached 65 institutions: almost a 100% increase (Table 4.1).

Table 4.1   USA Electric Cooperatives providing rural broadband services

Electric Cooperative	Year	Area	Services Provided	AE	GPON/EPON
Douglas Fast Net	2002	OR	Business Services, Data	X	
NineStar Connect (ex-Hancock Telecom)	2002	IN	Voice, Data, Video, Smart Grid		X
Blue Ridge Mountain Electric Membership Cooperative	2006	GA-NC	Data	X	
Southeast Colorado Power Association	2009	CO	Data	X	X
Arrowhead Electric Cooperative (1)	2010	MN	Data, Voice		X
Habersham Electric Membership Cooperative (partner of Internet EMC)	2010	GA	Data, Business Services	X	
Kit Carson Electric Cooperative (1)	2010	NM-CO	Data, Voice, Smart Grid		X
Lumbee River Electric Membership Corp (1)	2010	NC	Smart Grid, Data, Video, Voice		X
NEXT (North Alabama Electric Cooperative) (1)	2010	AL	Voice, Data, Video	X	X
Plumas-Sierra Telecommunications (1)	2010	CA			
Ralls County Electric Cooperative (1)	2010	MO	Data, Smart Grid		
United Electric Cooperative (1)	2010	MO	Data, Video, Voice, Smart Grid	X	X
Co-Mo Electric Cooperative	2011	MO	Data, Voice, Smart Grid, Video		X
Enlite Fiber Optic Network (Consolidated Electric Cooperative)	2012	OH	Video, Data, Voice	X	

(Continued)

Table 4.1 (Continued)

Electric Cooperative	Year	Area	Services Provided	AE	GPON/EPON
Lake Region Electric Cooperative	2012	OK	Data, Voice		X
GCEC Telecom (Grayson-Collin Electric Cooperative)	2013	TX	Video, Voice	X	
Guadalupe Valley Electric Co-op (GVEC.net)	2013	TX	Data, Video, Voice	X	X
Midwest Connections (Midwest Energy Cooperative)	2013	MI	Data		X
BARC Electric Cooperative	2014	VA	Data		
Bolt Fiber Optic Services (Northeast Oklahoma Electric Cooperative) (1)	2014	OK	Data, Video, Voice		X
OPALCO (Rock Island Communications)	2014	WA	Voice, Data	X	
Barry Electric Cooperative	2015	MO	N/A		
Valley Electric Association, Inc. (VEA)	2015	NV	Voice, Data, Video		
ARIS (Arkansas Rural Internet Service)	2016	AR	Data, Video, Voice		
Carolina Connect (Mid-Carolina Electric Cooperative)	2016	SC	Data		X
Ciello (San Luis Valley Rural Electric Coop)	2016	CO	Business Services, Data, Voice		X
Sho-Me Technologies	2016	MO	N/A		
BEC Fiber (Bandera Electric Cooperative)	2017	TX	Data		
Callaway Electric Cooperative (Callabyte Technology)	2017	MO	Voice, Data, Video		
Continental Divide Electric Cooperative	2017	NM	Data		

(Continued)

Table 4.1   (Continued)

Electric Cooperative	Year	Area	Services Provided	AE	GPON/EPON
Gibson Connect (Gibson Electric Membership Corporation)	2017	TN	Data		
Jackson County REMC/Jackson Connect	2017	IN	Data		X
Maquoketa Valley Electric Cooperative	2017	IA	Voice, Data, Video		X
Mecklenburg Electric Cooperative	2017	VA	N/A		
Mille Lacs Energy Cooperative	2017	MN	N/A		
Ntera	2017	WI	Data		
NEXT (North Arkansas Electric Cooperative)	2017	AR	Data, Video, Voice		
Pemiscot-Dunklin Electric Cooperative	2017	MO	Data, Video, Voice		
Prince George Enterprises (Prince George Electric Cooperative)	2017	VA	Data		
Roanoke Electric Cooperative	2017	VA	Data		
SEMO Electric Cooperative	2017	MO	Voice, Video, Data		
Tombigbee Communications (Tombigbee Electric Cooperative)	2017	GA	N/A		
Tri-County Electric	2017	TN	N/A		
Volunteer Electric Cooperative	2017	TN	Data		
Central Virginia Electric Cooperative	2018	VA	Data		
Consolidated Cooperative	2018	OH	Data		
Craighead Electric Cooperative	2018	AR	Data, Smart Grid		

(*Continued*)

Table 4.1 (Continued)

Electric Cooperative	Year	Area	Services Provided	AE	GPON/EPON
Great Lakes Energy	2018	MI	Data, Voice		
Holston Electric Cooperative	2018	TN	Data, Video, Voice		
Middle Tennessee Electric Membership Corporation	2018	TN	Data, Video, Voice		
OEC Fiber (Oklahoma Electric Cooperative)	2018	OK	Data, Voice		
OEConnect (Otsego Electric Cooperative)	2018	NY	Data, Voice		
Orange County Fiber (Orange County Rural Electric Membership Cooperative)	2018	IN	Data, Voice		
SCI Fiber (South Central Indiana Rural Electric Membership Corp.)	2018	IN	Voice, Data		
South Central Connect (South Central Arkansas Electric Coooperative)	2018	AR	Data, Video, Voice		
Taylor Electric Cooperative	2018	TX	Data		
Tipmont Rural Electric Membership Corporation	2018	IN	Data, Video, Voice		

Source: FCC

In Argentina, although the first cooperative society operating a telephone network was founded by David Atwell as early as 1887, at the time there were no specific laws for such type of institutions and their legal ownership regime. Even with the approval of the Code of Commerce in 1889 where cooperatives were included as commercial societies, it was just in 1926 that a comprehensive legal regime was enacted by National Congress (Law 11388). Since then, "cooperatives" are considered exclusively non-profit societies. Currently there are 1.037 active, registered utility cooperatives operating in Argentina. They have been delivering electricity, telecommunications, water and other services in over 600 locations. In the telecommunications business, cooperatives got birth in the 1950s typically to bring telephony in small towns. As legacy, locally-headquartered firms, their business ventures are customized, nimble and open to social expectancies, with high responsiveness to technological changes and to emerging IT services (Goussal, 2005–2006).

The advent of experiences in the provision of rural broadband services via non-traditional operators such as utility cooperatives has prompted a wave of interest in the re-examination of their role as an effective tool to foster the expansion of Next Generation Technologies in unserved areas. Such interest arises from the fact that utilities are well positioned for entry in these areas because of their existing infrastructure, rights of way and consumer bases. In addition, telephone and electricity co-ops have been historic providers in rural areas of the EU, USA and many developing countries, and experienced operators in the most difficult and the less profitable market niches.

In the early theoretical literature, Henry Hansmann explained the existence of cooperatives in the context of ownership arrangements. A cooperative firm will exist where such form of society minimizes the total costs of transactions and ownership between the firm and all of its patrons, including "the costs of market transactions for those classes of patrons who are not the owners, plus the costs of ownership for the class of patrons who own the firm". The incentives for consumers to set up a cooperative and generate additional finance are related to the utility they can gain in terms of financial and non-financial incentives (Hansmann, 1988; Sadowski, 2014).

Freshwater (1998) focused on investments by rural electric and telephone cooperatives and their relationship with economic development in their service areas. Since cooperatives usually take a longer term perspective and plan to be long term shareholders in the business, their decisions on network expansion problems deserve particular attention because they are rather different than other equity partners focused in short term investments (like the sale of the business in a few years). Cooperatives use to stay rooted in the community, because their business provides an ongoing stream of income and because it stays within the local service area. Cowhey and Klimenko (1999) pointed out that "cooperatives represent a quick and unorthodox way of raising funds for building out rural networks" when scarcity of capital precludes developing regions from expanding telecom infrastructure in the rural areas. The point of raising funds was revisited by Mamouni et al. (2016).

Fairchild (2000) noted that electric utilities including rural cooperatives have added the retail high-speed access business to their existing dark fiber capacities, and electric co-ops partner with telephone co-ops to deploy high-speed access. Worstell (2002) focuses on the advantages of rural cooperatives to enter into the broadband business due to their direct relation to their established customer base and their suitability to perform more incremental installs. Furthermore, as soon as technology development easies convergence of electricity distribution and telecom access, new means of mixed-infrastructure could use the rights of way and their corresponding conduits (Samarajiva et al., 2002). The study on municipal electric utilities by Gillett et al. (2004) finds an apparent relationship among the willingness of local providers to enter the telecommunications market and the availability of communications infrastructure to support electric utility operations–i.e. SCADA or Smart-Grid networks-, due to scope economies (Gillett et al., 2004).

On the other hand, in de-regulated markets rural telephone co-ops of are facing increasing competition among national players and wireless Internet providers. In most rural areas, the inability to raise funding to update the infrastructure will prevent the local cooperatives to compete with the new technologies that may overtake traditional phone service (Lawyer, 2004). The investment behavior of independent telecommunications providers in the long term has been examined in historical studies, e.g. in USA by Milton Mueller (1991), both the Swedish and the U.S. systems by Wallsten (2003), and the less-known Norwegian dual telephone system of a century ago by Harald Rinde (2001).

In the case of Latin America, the regulatory aspects of convergence between providers of different networks were also addressed by Segura (2000), Ravina (2001), Calzada et al. (2005) and Pelegrini et al. (2017). In Argentina, the study on electrical cooperatives of the province of Buenos Aires by Liana Acosta (2001) examined a number of variables of the operation in a survey of 193 local cooperatives in that region. Among others, she focused on number of users (rural and urban), power consumption per household, staff size, number of employees and additional services brought to the customers. It is worth mentioning that some 29% of the cooperatives of the sample were as well telephone providers, 16% were ISP and 11% owned cable TV distribution networks. The joint provision of services as a way to reduce overlapping network capital and overhead costs is undoubtedly the main driven factor, as appropriately analyzed by Levi (1988), Falch (1996), Anderson and McCarthy (1999), Gillett et al. (2004), Mikami (2010) and Sadowski (2014).

4.3 The three Layers of Rural Broadband Feasibility

In one of its widest meanings, the word *feasibility* describes "the state or degree of being easily or conveniently done", thus resembling how difficult it is to do something. Rural broadband is a vast and complex arena for NGT projects, due to the juxtaposition of constraints and requirements of different nature, undefined scope and – often-antagonistic trends.

In the case of a proposed distribution of the service by utility cooperatives, *feasibility* may be analyzed along three layers, partially superimposed each other: intrinsic (related to the pure broadband distribution business), institutional (relative to the overall economic and financial status of the cooperative and eventual benefits of adding rural broadband distribution to its services portfolio) and customer (relative to the net benefits of adopting the service by prospective rural users).

Let's consider a few facts:

- Rural broadband is a networked service. Its consumption cannot take place in isolation, since the customer requires the existence (and access) to a network. Therefore, as in any public telecommunications service the demand exhibits a conditional, two-stage behaviour (access and use), implying the existence of externalities and critical mass. Prior to make *consumption* of any broadband service, the user needs to have *access* to the network.

- Likewise, as in any telecommunications service the provision of rural broadband is granted (and constrained) by the regulatory framework (licenses, competition regime, area of service, interconnection regime, tariffs, etc.). Furthermore, in order to be legally entitled to hold a broadband license, the cooperative has to be enabled by an institutional statement, and financially suitable to embark in the new business based on available capital and loans.

- Beyond the assessment of economic feasibility at the side of the provider, in the case of a non-profit cooperative there is another requirement for the rural broadband service: it has to be feasible also from the side of the customer (which is typically, also a member of the institution). Evaluating the feasibility of the broadband consumption from the side of the customer implies the examination of the consumer surplus in a bottom-up approach. To a extent, such a wider sense of feasibility is related to social and economic development changes (eventually, benefits) produced as a result of the use of broadband services.

In our view, the successful addition of broadband services by rural utility cooperatives requires a mandatory assessment of feasibility along these three layers. In other words, the proposed rural broadband service has to be intrinsically sustainable, but also has to add a net benefit to both the institution and its members/customers.

4.3.1 Intrinsic Feasibility Analysis

An obvious requirement imposed to any provider before launching a broadband business would be its *intrinsic economic feasibility*. A service is said to be feasible if it yields positive net returns (i.e. discounted profits exceed discounted costs). Then, the Net Present Value has to be positive, the benefit/cost ratio must be 1 or greater, and the Internal Rate of Return must exceed the envisaged discount rate (Workman & Tanaka, 1991).

Intrinsic feasibility -from the side of the provider- results from the contingent combination of two *local* patterns: supply (mostly, up to the provider) and demand (mostly, up to the customer). Both sides represent a particular bundle of unique, legacy characteristics of the envisaged provision- and consequently, hard to replicate or to generalize in different contexts-. Naturally, the treatment of intrinsic feasibility factors does not ignore their overlapping with the other two layers: *institutional feasibility* and the *customer* (demand) *feasibility*. Such interactions vary in scope and strength according to the contingent situation of each utility cooperative, facing the broadband service.

The most "independent" analysis of intrinsic feasibility would be that of Group "D" cooperatives, i.e. public service institutions such as farmers associations that do not own yet any network, but whose members wish to provide broadband connectivity themselves in underserved or unserved areas. This case is much like starting the business from scratch, with the highest degree of freedom respect to society forms, capital needs and operational characteristics (including joint-ventures with cooperatives and/or public-private partnerships, leasing of antenna heights and rights of way, subcontracts with government offices, municipal networks and non-government organizations). In any case and regardless of the envisaged level of operational independence the new "business unit" may benefit from the existing membership base and institutional organization, local headquarters, trained staff, access to available equity/credit sources, and other intangible assets. Conversely, it may also suffer from accumulated liabilities and debt stocks, institutional drawbacks and scarcity of capital and cashflow margins.

The examination of the demand pattern at this layer usually requires a battery of customer surveys or geo-referenced studies of income and e-skillness groups, willingness-to-pay, desired levels of Quality of Service/Quality of Experience (QoS/QoE) and ultimately, a realistic forecast of annual adoption ratios in a spatial blueprint, so as to prepare a deployment plan tailored to the designated service area with sequences of spatial and temporal phases, rooted on accurate data of appropriate granularity and fine-tuning (Goussal, 2017).

The supply pattern, at the core feasibility layer is the expression of a continuous tradeoff between tariffs and Operational Expenditures (OPEX), provided that any suitable combination of financial sources (equity/debt) will cater for the required sequence of Capital Expenditures (CAPEX). In any rural broadband project, CAPEX is the most difficult problem; it usually imposes an absolute limit for network expansion initiatives undertaken by utility cooperatives. This is because, as opposed to private providers, the operation of a non-profit utility is not supposed to generate -from its existing business portfolio- a continuous accumulation of revenues in the form of a capital upgrades to be allocated to new investment ventures. The required CAPEX will be thereby limited to the amount of membership equity accrued via e.g. one-time subscription fee, and debt from investment loans either by commercial or cooperative banks and public funding through ad-hoc broadband programs.

The second supply limit is the costs pattern. Again here, cooperatives of the "D" group will have the widest autonomy in terms of operation cost decisions for their proposed broadband ventures, though associated to the highest degree of uncertainty and risk.

Engineering Views

Technological solutions for providing broadband access in rural areas, either by wireless, optical or mixed transmission alternatives are in general, considerably more expensive than those utilized in urban areas, as a result of the lower rural population density. Rural zones in USA have 72% of the land area but hold just 14% of the population. An analysis of utility cooperatives in USA with existing infrastructure in their rural service areas (such as electric distribution lines) shows a clear trend to go as far as possible with two optical technologies: GPON (Gigabit Passive Optical Networks) and AE (Active Ethernet). In 2015, 19 out of 33 cooperatives providing rural broadband distribution in USA reported the use of any optical alternative, and in late 2016 residential access with gigabit capacity was offered by 87 of them (Trostle & Mitchell, 2017).

Today, the use of GPON seems to be the starting technology in the provision of access in suburban and urban districts, and the preferred option for backhaul links connecting rural nodes. The reason is simple: investing in passive platforms is a sort of "future-proof" expansion strategy, destined to minimize the cost of successive capacity upgrades and technology changes. As soon as the NGT applications and broadband users will demand more capacity, the optical distribution platform will remain ready to provide it for many years. Upgrades will take place only in the electronics and active equipment at both ends of the FTTH network: the Optical Line Terminal at the central office of the provider (OLT) and the Optical Network Unit at the customer premises (ONU). In USA, electric cooperatives with ongoing rural broadband projects in most cases have chosen any type of optical transmission technology (e.g. GPON, Active Ethernet) due to their acceptable compromise among highest capacity, best transmission quality and lower engineering costs (combination of operational costs and capital costs in the long term) (Tables 4.1 and 4.2).

However, such FTTH model is far from a representing a generalized option, since there are legacy constraints and conditioning factors affecting the engineering design and the deployment strategy. In Argentina, most cooperatives providing rural broadband connectivity (group "A" sample) have started (and still are) using wireless technologies in last mile and even in backhaul segments, either from proprietary or normalized standards. In most cases, the older systems are in good operating conditions and still have spare capacity, thus the best time for substitution by optical transmission equipment has to be carefully evaluated. Telephony cooperatives were the first Internet providers in rural areas. Since they do not own physical lines and poles, the expansion strategy relied mainly on cheap, proprietary-standard, fast-to-deploy wireless equipment -at the expense of their limited capacity, rapid technology obsolescence and higher maintenance costs-. In some cases, they have wireless infrastructure still working in backhaul connections, because the central office is based in a town far from the nearest optical backbone node (Joseph et al., 2018).

Electric cooperatives in Argentina started operating in the broadband distribution business around 2000–2001, where a key change in the regulatory framework enabled them to apply for telecom licenses and value-added services. In general, their choice was to start serving urban districts of their designated service areas, including medium-size cities where naturally, they have to compete with large incumbent telecom operators by using state-of-the-art optical or Hybrid Fiber-Coaxial (HFC) technologies. The

Table 4.2   Other USA Electric Cooperatives providing rural broadband services

Electric Cooperative	Area	Services Provided	AE	GPON/EPON
Communications Access Cooperative (CACHE, Hood River Electric Cooperative)	OR	N/A		
ConnectAnza (Anza Electrical Cooperative)	CA	Voice, Data, Video		
Delaware County Broadband Initiative (DCBI)	DE	N/A		
Elevate Fiber (Delta-Montrose Electric Association, DMEA)	CO	Data, Voice		X
French Broad Electric Membership Corporation	NC	Data		
IllinoisNet.com (Illinois Electric Cooperative)	IL	Voice, Data, Video		
Mescalero Apache Telephone (1)	NM	N/A		
OzarksGo	AR-OK	Voice, Data, Video, Smart Grid		X

Source: FCC

distance to the available interconnection points, the high price and scarcity of broadband capacity as offered by then by the incumbents forced utility cooperatives to join themselves to build their own backhaul networks, as occurred in 2007 with a group of ten cooperatives that partnered with the public electric company of the province of Cordoba (EPEC) to connect themselves by using a 300-km optical transport infrastructure over the rights of way of EPEC in a High Voltage Line.

Notwithstanding, the introduction of rural connectivity by electric cooperatives has been a much slower process, where only the wealthiest and the bigger institutions have extended their networks beyond the urban limits. Among the differences with their sisters in USA, the smaller amount of investments and soft loans available from public sources or universal access funds via competitive bidding is one of the conditioning factors in the selection of the right deployment strategy.

Other constraints from the point of view of the utility cooperative are:

- The spatial location pattern of targeted subscribers (number, density, income range, distances).
- The matrix of local vs. external costs (i.e. ability of utility staff to install/maintain FO networks).
- The location, available capacities and cost matrix for the backhaul options.

A prototyping cost-conscious design, although with enough capacity to run NGT applications and to cater for broadband expected growth, with deployment along existing rural distribution lines and poles of an electric cooperative, can be summarized as follows:

- Gigabit Passive optical network (GPON) as far as possible, without amplifiers or regeneration.
- Overbuilt, aerial installation with Self-Supported-All-Dielectric optical cable (ADSS) along the existing rights of way of the power lines.
- Installation, operation and maintenance to be performed by the utility staff (training required).

Because of the distances involved and the low density of subscribers in rural areas, the laying of optical cables (distribution and drop trenches) is the largest cost component for GPON networks. In order to get a realistic matrix of costs for engineering analysis in Argentina and Latin American countries, we gathered data on the spatial distribution of lines, user densities and costs of a number of rural electrification projects from bidding documents, along with the blueprint of distribution lines, the length and number of users from a sample of rural electric cooperatives not yet providing broadband services (group "C"). Our findings can be summarized as follows:

- Rural electric networks in projects financed from federal programs of the Secretary of Energy of the Nation (PROSAP) and other public sources in Argentina, show user densities typically between 0,59 and 2,11 households/Km (average: 1,16 households/Km) Though there are densely populated rural areas like the province of Misiones with values up to 6,57 households/Km.
- Typical rural networks of electric cooperatives in the north of the country show values of about 1 user/Km. There are single users up to 70 Km away from the main distribution line, but the "native" GPON would be technically bounded to its maximum distance (20 Km).

- The good condition of utility poles is a critical constraint affecting the prototyping design (assumption II). Poles in lines beyond 10 years old or in zones affected by floods or storms will have frequent collapses and may be not suitable for laying ADSS cables. Utilities with large distribution areas have to replace as usual, more than 500 wooden poles a year.
- As an example of the prototyping design, we computed the matrix of costs for a 100-Km GPON network with the average of an extremely-low density of rural power lines (only 116 prospective rural users). The cost breakdown is depicted in Figure 4.1. As expected, optical cables and installation work is by far the greatest cost component accounting for 81,4% of the total. Laying aerial ADSS and drop cables with own staff in cooperatives of Argentina is 3 to 4 times cheaper than in USA, although this is the largest cost component (46,9%).
- The cost per subscriber in the prototyping design is directly and heavily influenced by the subscriber density, and may be almost halved for the maximum real density (2,11/Km).
- The total cost of overbuilding a 100-Km Rural GPON network in Argentina is about five times cheaper than that of the rural electric distribution line (without optical infrastructure).

Not only the cost per subscriber, but also the network architecture and technology will be therefore, highly related to the spatial location pattern of users (constraint 1). Furthermore, there are two sites of utmost importance: the location of the Optical Line Terminal (OLT), that determines which and how many urban and rural subscribers could be reached within the GPON

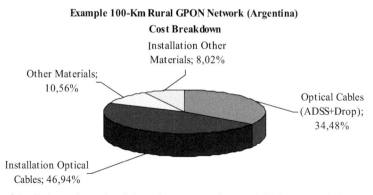

Figure 4.1   Estimated cost breakdown in a prototyping rural GPON network in Argentina.

maximum span (20 Km) and the available point of interconnection with the optical trunk network, which will determine the distance and the cost of the backhaul link. In the most favourable case, the rights of way of power lines can be used for both purposes (backhaul and distribution), even with a common optical cable.

When the bounds of the rural area are beyond the maximum reach of classic, "pure" GPON, there are other possibilities to extend the coverage without sacrifice of capacity, namely:

- Use of extended differential reach systems (EDR GPON), as standardized in Rec. ITU-T 984.7, based in the use of optical components suitable to maintain power levels enough to support attenuation ranges of 13–28 dB (Class B+) to 17–32 dB (Class C+) consistent with maximum physical distances of 40 to 60 Km.
- Use of high-count ADSS cables with dedicated strands (without splitters) for distant customers.
- Use of optical range extenders with pole-mounted, outdoor enclosures.
- Use of mixed GPON/Active Ethernet architecture with compatible OLT/Optical Distribution Network (ODN) devices.

The cost of backhaul links depend on the location of the node of the nearest optical backbone. As long as possible, the use of electric poles in the rights of way of the utility cooperative is an economical way to deploy in a single optical cable with different fiber strands the links for both purposes distribution and backhaul, besides reducing overall operation and maintenance costs. In the case of telephony cooperatives, or when backbone nodes are out of a power distribution line, some institutions own, rent or share point-to-point microwave links. In Argentina, where optical backbones in several regions are far away, this was the choice of telephony cooperatives in nearby locations grouped themselves to share capacity and costs of a common long-haul microwave link.

4.3.2 Institutional Feasibility

The ability to undertake a rural broadband project by a utility cooperative requires an examination of data from its business operation in the current portfolio of services provided. Such is the purpose of the feasibility analysis at the institutional layer: an assessment of its performance as a public service provider in the last years. In this vein, data from balance sheets are regarded as a reliable source of data at the public level: they are supposed to have been prepared, approved by the board of directors and independently audited by a certified accountant. A single balance sheet is however, just a snapshot taken at the end of the reporting period of the utility, and cannot reflect its whole operation behaviour. The institutional feasibility analysis should better rely on a *series of balance sheets* covering the latest audited periods, even from non-consecutive years. Such methodology was utilized in the study of utility cooperatives in Argentina, aimed at identifying and evaluating aspects of the institutional feasibility of adding rural broadband distribution as a new provision in their designated service areas. Balance sheets are audited documents containing key information about three main aspects: 1) Statement of assets, liabilities and equity, 2) Statement of income, operating revenues and expenses, and 3) Cash flow statement.

Furthermore, the relative wealth of a cooperative in its business operations is reflected as in any other utility, by measuring economic and financial ratios (debt, liquidity, profitability, etc.). The sample of 74 utility cooperatives whose balance sheet series were studied is listed in Table 4.3.

Cooperatives with rural coverage of electricity, gas or clean water networks but without broadband services (group "C") or those ones providing just urban broadband connectivity (group "B") show costs structures where the main cost components are salaries, broadband costs and depreciation of assets. An example of actual values is the comparison among two real patterns of Fully Allocated Costs (FAC) in a large "group B" utility cooperative of Argentina: that of the entire portfolio of electricity, social services and urban connectivity, and that of just the broadband distribution (Figures 4.2 and 4.3).

In this case, while salaries, taxes, commissions, and other costs show similar weights, there are two differences: a) the depreciation of broadband assets is greater than that of the total services (13% vs. 2%) and, b) the rest of pure operational costs (OPEX) on connectivity services, are proportionally lower than those from the entire portfolio provided by the utility cooperative. Thus revealing that -among other components- the cost of trunk connectivity, although still high and often purchased in Argentina from two or more providers due to the limited capacities offered, is not anymore the main barrier in the seeking of intrinsic feasibility. A closer examination of the institutions in the sample and the differences in the relative weight of the main cost components reveals that the cost of bulk broadband and backhaul is proportionally greater for the smallest institutions and/or for those where broadband is one of the main provisions, such as telephony cooperatives. Salaries and overhead costs (e.g. management, sales/marketing and financing) are proportionally greater in large institutions where broadband is just a small fraction of their total sales and where electricity distribution is far the most important business.

Table 4.3   Sample of utility cooperatives in Argentina with data recorded from balance sheet series

Name of the Cooperative	Province	Group	Name of the Cooperative	Province	Group
ARMSTRONG	SANTA FE	A	MARIA SUSANA	SANTA FE	B
CAMET	BUENOS AIRES	A	MONTECARLO	MISIONES	B
CONCORDIA	ENTRE RIOS	A	MOQUEHUA	BUENOS AIRES	B
CORONEL CHARLONE	BUENOS AIRES	A	PINAMAR	BUENOS AIRES	B
CORONEL DU GRATY	CHACO	A	POZO DEL MOLLE	CORDOBA	B
DE LA GARMA	BUENOS AIRES	A	RIO COLORADO	LA PAMPA	B
EL CALAFATE	SANTA CRUZ	A	ROJAS	BUENOS AIRES	B
MONTE	BUENOS AIRES	A	SAN MARCOS SIERRAS	CORDOBA	B
ONCATIVO	CORDOBA	A	SANTA ROSA	LA PAMPA	B
OBERA	MISIONES	A	SANTA ROSA DE CALAMU-CHITA	CORDOBA	B
PUERTO RICO	MISIONES	A	TORTUGUITAS	BUENOS AIRES	B
RAMALLO	BUENOS AIRES	A	VILLA HUIDOBRO	CORDOBA	B
REALICO	LA PAMPA	A	VILLA SANTA ROSA	CORDOBA	B
SAN ANTONIO DE ARECO	BUENOS AIRES	A	WHEELWRIGHT	SANTA FE	B
SAN BERNARDO	CHACO	A	ZARATE	BUENOS AIRES	B
SAN MANUEL	BUENOS AIRES	A	16 DE OCTUBRE	CHUBUT	C
SAN MARTIN DE LOS ANDES	NEUQUEN	A	ADOLFO ALSINA-CARHUE	BUENOS AIRES	C

(*Continued*)

Table 4.3   (Continued)

Name of the Cooperative	Province	Group	Name of the Cooperative	Province	Group
SANTA SYLVINA TELEF	CHACO	A	AVELLANEDA	SANTA FE	C
SUIPACHA	BUENOS AIRES	A	CORZUELA	CHACO	C
VILLA GENERAL BELGRANO	CORDOBA	A	CHARATA – 212	CHACO	C
AZUL	BUENOS AIRES	B	CHARATA – 27	CHACO	C
CARLOS PAZ	CORDOBA	B	HERMOSO CAMPO	CHACO	C
COLON	BUENOS AIRES	B	JUAN JOSE CASTELLI	CHACO	C
CURARU	BUENOS AIRES	B	LAS BREÑAS	CHACO	C
GENERAL PICO	LA PAMPA	B	MACHAGAI	CHACO	C
HUINCA RENANCO	LA PAMPA	B	PEHUAJO	BUENOS AIRES	C
JESUS MARIA/C. CAROYA	CORDOBA	B	PERGAMINO	BUENOS AIRES	C
JOVITA	CORDOBA	B	PRESIDENCIA DE LA PLAZA	CHACO	C
LAGUNA LARGA	CORDOBA	B	RIO CEBALLOS	CORDOBA	C
LAS ACEQUIAS	CORDOBA	B	ROQUE SAENZ PEÑA	CHACO	C
LAS FLORES	BUENOS AIRES	B	SAN PEDRO	BUENOS AIRES	C
LAS HIGUERAS	CORDOBA	B	SANTA SYLVINA AGROP	CHACO	C
LAS PERDICES	CORDOBA	B	TRES ISLETAS	CHACO	C

(Continued)

Table 4.3 (Continued)

Name of the Cooperative	Province	Group	Name of the Cooperative	Province	Group
LOS CONDORES	CORDOBA	B	VILLA ANGELA	CHACO	C
LUJAN	BUENOS AIRES	B	VILLA BERTHET ELECTR	CHACO	C
MAIPU	BUENOS AIRES	B	GENERAL PIRAN	BUENOS AIRES	C
MARIANO MORENO	BUENOS AIRES	B	TRELEW	CHUBUT	C

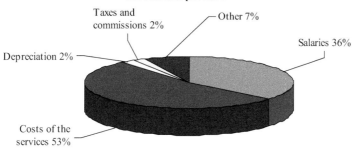

Figure 4.2 Costs of all services of a typical utility cooperative providing urban broadband (Argentina-Group B).

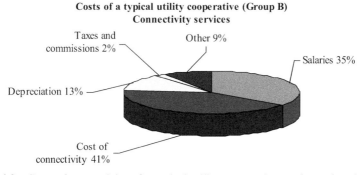

Figure 4.3 Costs of connectivity of a typical utility cooperative serving only urban areas (Argentina-Group B).

Concerning utility cooperatives already providing broadband access in rural areas (group "A"), in a sample of 58 institutions of varying size and location whose data were extracted, digitized and re-calculated by de-aggregating bulk data contained in series of balance sheets covering at least one of the latest 3 years of operation (2015–2017), we extracted and compared the values of net profits or losses for broadband service and for the other main service provided (Electricity distribution for electric utilities and Telephony for telephony co-ops). In similar way, we computed the net profits or losses of broadband access for cooperatives providing it only in urban areas, against the values for the other main service provided (Electricity distribution). The results are shown in Figures 4.4 and 4.5.

From the sample of 19 cooperatives in Group "A" and 29 in Group "B", no major differences arise among both patterns, and the analysis confirmed the following facts:

- Broadband distribution does not have to be (inherently) more or less profitable than other networked services provided by the utility. Taking both groups together, 25 out of 58 utilities made net profits from broadband, while 23 did so from electricity or telephony.

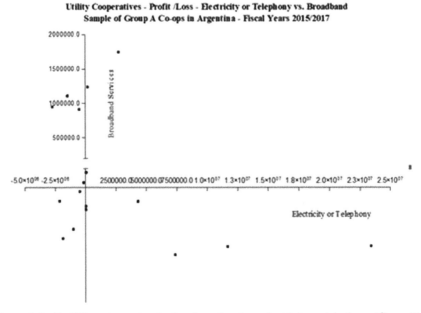

Figure 4.4   Profit/Loss in rural and urban broadband vs. electricity or telephony (Group "A" utilities).

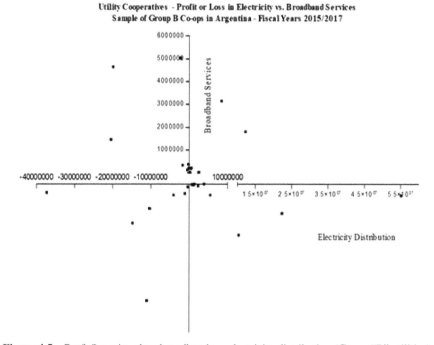

Figure 4.5 Profit/Loss in urban broadband vs. electricity distribution (Group "B" utilities).

- Profitability of broadband distribution is not (inherently) related to profitability of the other services delivered by the utility: only 9 out of 58 institutions obtained net profits in both broadband and the main other service (3 in Group A and 6 in Group B).
- Curiously, the total sum of net results of the 58 institutions yields an overall profit in both the main service and broadband, but with opposite behaviour in the services by each group. In Group B, the total sum of 29 utilities yields a loss in broadband services and a net profit in electricity, while in Group A the profit was in the sum of broadband results and the loss was in the other main service (electricity or telephony).
- Providing just urban broadband (Group B) or both rural and urban broadband (Group A) -according to the proportion of users and possibilities of each institution-, does not turn more or less profitable the service but in a minimal part (3 out of 19 profitable cases in Group A and 6 out of 29 in Group B)

- There is a large proportion of cooperatives with losses in both provisions e.g. electricity and broadband distribution in the same year (24 cases 10 in Group A and 14 in Group B-)
- A large net profit in Electricity in a single fiscal year can add a capital source, but does not ensure profitability in an envisaged broadband distribution project (cases beyond middle of the right X-axis in Figures 4.4 and 4.5)
- Most of the sampled institutions of either group, either large or small in membership or assets already providing broadband services, show along years average profits or losses of small proportion (points near the origin in X and Y axes).

4.3.3 The Third Layer: Willingness-to-Pay vs. Consumer Surplus

In the conventional wisdom of feasibility in telecommunications services, the business has to be profitable for the provider, as analyzed so far from the supply side. The other side is the demand, where usually we refer to concepts such as adoption, take-up rates, willingness-to-pay and prices. An operator needs to carefully check how many households passed will subscribe at a given price to each quality level of the broadband portfolio offered, how long will last adoption in each type of consumer group (business, residential, government, etc), and what their willingness-to-pay will be.

Demand studies for broadband in rural areas include such aspects and social/economic effects, geographical differences, pricing and efficiency of public funding support (Whitacre et al., 2012, 2014, 2015; Carare et al., 2015; Van Der Wee et al., 2015; Lee et al., 2017; Pant et al., 2017).

However, when providers are utility cooperatives, feasibility requires a third layer of analysis: the demand side (consumers). This is because in utility cooperatives, the consumers -to a great extent- are as well members (owners) of the institution. Therefore, the provision of rural broadband has to be feasible from the supply side but also from the demand side, because it does not make sense to start a business feasible for the cooperative, but without translation into net benefits to its members/consumers. Measuring benefits to them means to determine the implied value of broadband by considering what amount a consumer would be willing to pay for service, compared to what he is actually paying. This is often referred by economists as *consumer surplus* ((Rembert et al., 2017).

The focus in this layer should be on assessing the consumer surplus vis-a-vis the pricing and the quality attributes of the broadband services offered.

This implies finding the true value/benefit of using a rural broadband access by calculating the cost of alternative ways, otherwise available to the user to retrieve or to send information, i.e. the cost of travel and wages foregone for which the broadband service substitutes), minus the total price paid for using the service. Several studies have estimated the values in different countries, geographical areas and qualities of service. For example in Australia, a choice modelling study found that a household taking up a plan of 100 Mbs download/40 Mbs upload would have an average benefit of $59.10 per month relative to average speeds at the time of the survey. In USA, for a Gigabit access with unlimited usage priced at a flat rate of $100, the monthly average consumer surplus was estimated at $280 (Nevo et al., 2013).

A study on broadband un-adopters by Whitacre & Rhinesmith (2016) found that cost was the main reason for non-use. No- need and inadequate computers ranked second and third reasons. Findings reveal that un-adopters with income levels below $40.000 would benefit from broadband subsidy plans, since those low-income groups have above-average rates of un-adoption. Such income level translates into a $100 a month limit in order to keep the telecommunications expenditures at no more than 3% of the household income. Obviously, this limit will be lower in a developing country and likely in rural areas, which suggests that cooperatives have to set up pricing policies in line with the actual ranges of household income in their respective service areas. Otherwise, the take-up rate will be too low and the planned rural broadband project will be unfeasible. However, different user groups with similar income levels may have too different values of consumer surplus and then, varying amounts of willingness to pay. Rural residents that have to commute often to the city will likely find more value in the broadband access provided by their cooperative (depending on price).

A formal analysis of consumer surplus, access/membership policies and feasibility of investments undertaken by cooperatives was that of Rey and Tirole (2007). The authors concluded that non-discriminatory consumer cooperatives are highly fragile institutions, because new members free ride on the investment of established members, with the risk of underinvestment (preventing the cooperative to go ahead with e.g. a rural broadband plan). This issue is handled by many utilities with a differential pricing policy, e.g. requiring one-time capital contributions from new members as a way to add funding in a planned expansion of the broadband network.

It is important to distinguish the concepts of customer surplus and willingness-to-pay. The study by Petersen (2017) used a large data set of over 25,000 consumer broadband plans in the 29 largest broadband markets between January 2010 and November 2012. Data included variations in market structure and types of broadband plans, allowing for estimation of willingness to pay for upgrades in the service quality. He concluded that consumers were willing to pay $75.75 more for speeds of 25 Mbps – 50 Mbps than slow plans, for plans that meet the FCC 2015 minimum speeds that can be called broadband (25 Mbs download/4 Mbs upload). But the problem is not with the maximum speed of access: even for the next 5G where the connection quality is expected to be very high, the strongest constraint for providers will be the willingness to pay of user groups (Hansen, 2017).

Finally, and beyond the calculation of consumer surplus for prospective user groups, in the case of USA and other countries with public funding for rural broadband plans indirect benefits are as well important. A study on the effects of the expansion of broadband access in 1999 to 2007 on labour market outcomes in USA found positive significant correlations between the access to broadband services and employment rates, especially in rural and isolated areas. The access to broadband in a new county is associated with approximately 1,8% increase in employment rates with larger effects in rural and isolated areas (Atasoy, 2013; Van der Wee et al., 2015; Draca et al., 2018).

In sum, the ultimate limit in every rural broadband project undertaken by utility cooperatives is the third layer: the feasibility from the side of the customer – which is likely also a member and owner of the institution-. The household income is an important threshold for the *ability to subscribe*, but is not per se a trigger for the willingness to pay. The customer surplus is more important for rural user groups as a driver of adoption. In the case of rural areas with low income households, public funding from universal access or rural broadband programs may overcome the gap between the values of intrinsic and customer feasibility layers. The appropriate estimation of indirect benefits, in particular savings in the cost of public services and e-government is a key tool for an efficient assignment of funding to utility cooperatives via competitive bidding.

4.4 Conclusions

Utility cooperatives are historic providers of networked services in several countries, with deep roots in small towns and rural areas. Their genetic values -such as member-ownership and strong community sense- nowadays cohabitate with full professional management, commercial and engineering practices. Broadband distribution and NGT applications are natural extensions of their service portfolios. Nevertheless, even in industrialized nations, they use to be financially weak, and consequently unable to embark in infrastructure investments without public aid (ad-hoc regulation, access to trunk connectivity, soft loans, etc).

In this chapter we have discussed a three-layer framework for the analysis of feasibility in utility cooperatives actually providing rural broadband services or suitable to provide it under favourable conditions. Real data was obtained from a baseline study in Argentina by selecting, updating and cross-checking detailed information of 74 utility cooperatives nationwide, with available historical information at the micro-level retrieved on a one-by-one basis, including digitalisation and analysis of more than 150 annual balance sheets organized in series of 2 to 12 annual reports available from each entity between years 2006 to 2017.

Preliminary results show that any networked service delivered to rural areas by utility cooperatives is hard to turn into a profitable business, although in Argentina the strongly regulated fares of basic telephony and electricity have been a key constraint. Investment, financing and asset management strategies in each service are different. Electricity distribution in Argentina has a specific public fund to finance expansion of power lines in new areas, plus the public investment in trunk and regional lines, but consumption fares are strictly regulated. Conversely, broadband access enjoys limited public funding, but as a value-added service the pricing policy is essentially free. In general, soft loans or subsidies in Argentina available to utility cooperatives from universal access fund via competitive bidding are small and cannot be used for new broadband networks or expansions, but instead for technology upgrades and replacement of outdated distribution infrastructures. In USA, utility cooperatives applying to public funding and soft loans from federal broadband programs (FCC, etc) must include in the portfolio a basic quality access (25 Mbps download/4 Mbps upload) at a maximum price of $59,99 (Schmit and Severson, 2017).

In order to keep track of capacity growth and increasing demand, intrinsic feasibility of new rural networks require a cost-conscious deployment plan where passive optical technologies have a key role as "future-proof" investments, implying a high cost asset, although with long depreciation schedules. However, almost all institutions of Group "A" already providing rural access are currently using wireless technologies of limited capacity and proprietary standards. Comparison analysis of technology investment and feasibility strategies for rural broadband in utility cooperatives of Argentina and USA reveal that at the start, the former tend to give more priority to geographical coverage by selecting fast deployment, wireless technologies of limited capacity and modest reliability, geared to lower both CAPEX and service fares at the expense of capacity complaints, more frequent equipment changes and difficult upgrades in legacy equipment. In USA, rural co-ops seem to privilege sunk capital investment (such as buried optical cables), geared to high capacity, high reliability FTTH/FTTC assets with the aim of providing top quality, well priced access services while lowering OPEX and technology upgrades.

Regarding the Institutional feasibility layer, a sample of 58 utility cooperatives of Groups "A" and "B" already providing broadband services revealed that broadband distribution does not have to be (inherently) more or less profitable than other networked services provided by the utility. Less than half of them made annual profits in broadband (25) and other main service, electricity or telephony (23). Profitability of broadband distribution is not inherently related to profitability of the other services, and providing just urban broadband (Group B) does not turn more or less profitable but in a minimal part respect to the provision to both rural and urban areas (Group A), according to the proportion of users and possibilities of each institution-. The majority of institutions either large or small -in membership or assets- already providing broadband services, show along years average profits or losses of small proportion, thereby working with operating margins near zero. This is the main barrier from the institutional layer to raise funds on their own for rural broadband projects.

Finally, the analysis at the third layer –feasibility of the rural broadband service from the side of the demand- focuses on the estimation of the consumer surplus, departing from the particular case of utility cooperatives where to a great extent, the customers are also members of the institution. Therefore, it would be useless to add a new service, even when profitable at the institutional layer but without translation into net benefits to the consumers. Tariff levels in value-added services are relatively free for ISPs in Argentina, and then the first reaction in a monopoly market (the rural service area) would be to set prices in line with the willingness-to-pay. However, such policy would lead to un-adoption behaviour in low income user groups, or to null/negative consumer surplus in households with broadband expenditures beyond the limits of their income threshold. In such case, public funding or subsidies from universal access funds may contribute to overcome the gap between the intrinsic and the customer feasibility layers. This requires appropriate estimation of indirect benefits, in particular savings in the cost of public services and e-government, as well as local impacts in rural labour markets and employment rates.

References

Anderson, K. and McCarthy, A. (1999). Transmission pricing and expansion methodology: lessons from Argentina. *Utilities Policy* 8, pp. 199–211.

Atasoy, H. (2013). The Effects of Broadband Internet Expansion on Labor Market Outcomes. *ILR Review*, 66(2): 315–345.

Calzada, J. and Dávalos, A. (2005). Cooperatives in Bolivia: Customer ownership of the local loop. *Telecommunications Policy* 29, 387–407.

Carare, O., Mc Govern, C., Noriega, R. and Schwarz, J. (2015). The willingness to pay for broadband of non-adopters in the US: Estimates from a multi-state survey. *Information Economics and Policy* 30, 19–35.

Cowhey, P. and Klimenko, M. (1999). *The WTO Agreement and Telecommunications Policy Reforms a Report for the World Bank*. University of California at San Diego, USA.

Department of Communications and the Arts, Australia (2014) Independent cost-benefit analysis of broadband and review of regulation. Government of Australia.

Draca, M., Martin, R. and Sanchis-Guarner, R. (2018). *The Evolving Role of ICT in the Economy-A Report by LSE Consulting for Huawei* – The London School of Economics and Political Science UK.

Duvall, W. (2016). *The Evolving Role of Electric Cooperatives in Economic Development: A Case Study of Owen Electric Cooperative and Jackson Energy Cooperative* – Theses and Dissertations, Community and Leadership Development, University of Kentucky (UKnowledge), Paper 20.

Fairchild, D. (2001). *Infrastructure, Deregulation and Universal Access*. University of Florida, USA.

Falch, M. (1996). Cost and Demand for Telecoms Networks VIII ITS European Conference, Vienna, Austria.

Freshwater, D. (1998). *Utility Cooperatives as a Source of Equity Finance: Montana Examples*. RUPRI – Rural Policy Research Institute. University of Missouri. Columbia, Missouri-USA.

Gillett, S., Lehr, W. H. and Osorio, C. (2004). Local government broadband initiatives. Telecommunications Policy, 28(7–8), 537–558.

Goussal, D. (2005). Factores de correlación y divergencia en la expansión de la red telefónica fija en áreas servidas por cooperativas – Reunión de Comunicaciones Científicas y Tecnológicas Secretaría de Ciencia y Técnica-UNNE. http://www.unne.edu.ar/unnevieja/Web/cyt/com2005/7-Tecnologia/T-046.pdf

Goussal, D. (2006). *Conductas comparativas de expansión del acceso en redes cooperativas de distribución eléctrica y telefónica.* Reunión de Comunicaciones Científicas y Tecnológicas – Secretaría de Ciencia y Tècnica, UNNE. http://www.unne.edu.ar/unnevieja/Web/cyt/cyt2006/07-Tecnologicas/2006-T-050.pdf

Goussal, D. (2017). *Rural broadband in developing regions: alternative research agendas for the 5G era* – In: Skouby, K; Williams, I and Gyamfi, A. (Eds) Handbook on ICT Policies in Developing Countries – 5G Perspective – Vol 1, Wireless World Research Forum – River Publishers, Denmark. ISBN 978-87-93379-91-6.

Hansen, J. (2017). *5G and its Economic Aspects. Literature Review and Selection of a Connection Portfolio Under Risk* (Thesis dissertation) Norwegian University of Science and Technology. Oslo, Norway.

Hansmann, H. (1998). Ownership of the Firm. *Faculty Scholarship Series.* Nr. 5041.Yale Law School, Yale University USA.

Ida, T. and Horiguchi, Y. (2008). Consumer benefits of public services over FTTH in Japan: Comparative analysis of provincial and urban areas by using discrete choice experiments. *Information Society*, 1–17.

Joseph, J., Mukhopadhyay, A., Rudrapatna, A., Urrutia-Valdés, C. and Van Caenegem T. (2018). The Future of Fixed Access: A Techno-Economic Comparison of Wired and Wireless Options to Help MSO Decision Process. *SCTE-ISBE CableTec Expo – Fall Technical Forum.* Atlanta, USA.

Lawyer, K. M. (2004). *Bridging The Technological Gap: High-Speed Wireless Internet in Rural America.* M.A. Thesis Dissertation. Georgetown University, Washington D.C. USA.

Lee, H. and Whitacre, B. (2017). Estimating willingness-to-pay for broadband attributes among low-income consumers: Results from two FCC lifeline pilot projects. *Telecommunications Policy* 41, 769–780.

Mamouni, L. E., Watson, J., Mazzarol, T. and Soutar, G. (2016). Financial instruments and equity structures for raising capital in cooperatives. *Journal of Accounting & Organizational* Change, 12(1), 50–74.

Mikami, K. (2010). Capital procurement of a consumer cooperative: Role of the membership market. *Economic Systems* 34, 178–197.

Mueller, M. (1991). Telecommunications Development Models and Telephone History: a Response to the *World Bank ICTM News*, Vol. 2/2. University of Nebraska at Omaha, USA.

Nevo, A., Turner, J. and Williams, J. (2013). Usage-based pricing and demand for residential broadband, CSIO *Working Paper N 0121 Center for the Study of Industrial Organization at Northwestern Univ.*, Evanston, Ill.

Pant, L. and Odame, H. (2017). Broadband for a sustainable digital future of rural communities: A reflexive interactive assessment. *Journal of Rural Studies* 54, 435–450.

Pelegrini, M., Miranda, F., Cyrillo, I., Ribeiro, F., Ordonha, I., Martins, J., Pereira, E., Dutra, E. and Marques, M. (2017). Regulation for the Brazilian cooperatives (application of soft-regulation). *CIRED, Open Access Proc. J.*, 2017(1), pp. 2865–2868.

Peronard, J. and Just, F. (2011). User motivation for broadband a rural Danish study – *Telecommunications Policy* 35, 691–70.

Ravina, A. O. (2001). Cooperation Between Companies from Different Sectors in Argentina. *Annals of Public and Cooperative Economics*, 72(3), pp. 379–392.

Rembert, M., Feng, B. and Partridge, M. (2017). *Connecting the Dots of Ohio's Broadband Policy*. Swank Program In Rural-Urban Policy – Ohio State University, USA.

Rey, P. and Tirole, J. (2007). Financing and access in cooperatives. *International Journal of Industrial Organization*, 25(5).

Rinde, H. (2001). Knowledge Diffusion and Industrial Organisation: The Case of the Dual Norwegian Telephone System, 1890–1920 – *EBHA Conference 2001: European Business History Conference, Oslo,* Norway. 8–9.

Sadowski, B. (2014) Consumer cooperatives as new governance form: the case of the cooperatives in the broadband industry – *Working Paper No. 14-03 Eindhoven Centre for Innovation Studies (ECIS), School of Innovation Sciences – Eindhoven University of Technology* – The Netherlands.

Schmit, T. and Severson, R. (2017). *Exploring the Feasibility of a Rural Broadband Cooperatives in Northern New York*-EB 2017–05 – Charles Dyson School of Applied Economics – Cornell University, USA.

Schneir, J. and Xiong, Y. (2016). A cost study of fixed broadband access networks for rural areas. *Telecommunications Policy* 40 (8).

Segura, E. V. (2000). Convergence: Technical – Legal – Economical – Social and Political Obstacles for its Implementation from Argentina to the World. *Proc. ITS Biennial Conference-International Telecommunications Society, Buenos Aires* (Argentina).

Settles, C. (2014). Electric Co-ops Build Rural Broadband Networks. Broadband Communities.

Trostle, H. and Mitchell, C. (2017). Cooperatives Fiberize Rural America: A Trusted Model For The Internet Era in Community Networks – *Policy Brief. Institute for Local Self Reliance (ILSR)*.

Van der Wee, M., Verbrugge, S., Sadowski, B., Driesse, M. and Pickavet, M. (2015). Identifying and quantifying the indirect benefits of broadband networks for e-government and e-business: A bottom-up approach *Telecommunications Policy*, 39, 176–191.

Wallsten, S. (2003). Returning to Victorian Competition, Ownership, and Regulation: An Empirical study of European Telecommunications at the Turn of the 20th Century. *Working Papers Series, AEI-Brookings Joint Center for Regulatory Studies*. Washington DC, USA.

Whitacre, B. and Rhinesmith, C. (2016). Broadband un-adopters. *Telecommunications Policy* 40, 1–13.

Whitacre, B. Gallardo, R. and Strover, S. (2012). *Rural Broadband Availability and Adoption: Evidence, Policy Challenges and Options*. National Agricultural and Rural Development Policy Center, USA.

Whitacre, B., Gallardo, R. and Strover, S. (2014). Broadband's contribution to economic growth in rural areas: moving towards a causal relationship. *Telecommunications Policy*, 38, 1011–1023.

Whitacre, B., Strover, S. and Gallardo, R. (2015). How much does broadband infrastructure matter? Decomposing the Rural – Urban adoption gap with the help of the national broadband map. *Government Information Quarterly*, 32(3), 261–269.

Workman, J. and Tanaka, J. (1991). Economic feasibility and management considerations in range revegetation. *Journal of Range Management*, 44(6): 566–573.

Worstell, R. (2002). Cooperatives: a Workable Business Model for Broadband Implementation White Paper Series, Delta Enterprise Networks (DEN). Stuttgart, Arkansas (USA).

5

Blockchain, Trust and Elections: A Proof of Concept for the Ghanaian National Elections

Idongesit Williams and Samuel Agbesi

Aalborg University, Copenhagen, Denmark
idong@es.aau.dk, sa@es.aau.dk

The electoral process in most developing countries are often problematic. The root of the problem often stems from the lack of trust emanating from an event in the election process. Over the years, electronic voting solutions have been adopted. Unfortunately, these solutions have not solved the trust related issues on the electoral process. Blockchain is proposed in this chapter as a solution to this problem. This is because blockchain is a technology designed based on the trust by default principle. Persons who do not trust each other are able to facilitate transactions on blockchain in a trusted manner. The proposition is backed by a proof of concept using the Ghanaian electoral process as a case. This chapter concludes that blockchain is indeed the missing link in the facilitation of trust between the relevant stakeholders in a national, provincial or local election.

5.1 Introduction

This chapter is a proof of concept on how blockchain can be utilized in an election scenario in developing countries to facilitate trust between the responsible stakeholders. The general election (presidential and parliamentary elections) process in Ghana is used as a test case for the proof of concept. The focus of the proof of concept is not on the practical details of the technical dimensions of blockchain. However, it is on how blockchain as a technology can facilitate trust in an election process. The proof of concept is explained using scenarios that can be applied in Ghana. The scenarios outline the potential election process flow that could be facilitated using blockchain.

Why is this proof of concept necessary? Currently there are initiatives aimed at the utilization of blockchains to facilitate election in a handful of western countries. So far, blockchain trials have been conducted in minor municipal elections in Japan, USA, and Switzerland (Linver, 2018). It is being considered in Russia by an election watchdog (Huillet, 2018). However, no such trial has been conducted in developing countries. There was an attempt to conduct a trial in Sierra Leone in 2018, but it ended up being a practical proof of concept (Pollock, 2018). In reality, developing countries ought to explore the use of blockchains for their election processes. This is because elections in most developing countries are tense moments. In most cases, these tensions have resulted in either violence, prolonged legal litigation or even war. The underlying reason for these unsavoury events is the existence of mistrust in the election process. Either by the political parties, the candidates, or the electorate could harbour this mistrust. These three classes of stakeholders are central to the election process. Nevertheless, there is the need to adopt measures, including technological measures that would facilitate trust between these stakeholders. Blockchain is designed with the trust by default principle. This is why this technology is promoted in this chapter and used to develop a proof of concept on how it could be used to facilitate trust in an election process.

Currently, there are various technologies that facilitate Electronic Voting (e-voting). Nevertheless, these technologies are not designed with the trust-by-default principle. Based on experience from countries where these technologies are deployed, these technologies also elicit some level of mistrust. In the 2000 US presidential elections, the then Vice President Al Gore requested manual recount in Florida when the Gore campaign did not trust the outcome of the manual vote recount. In the 2018 Florida gubernatorial elections, machine and manual recount were adopted by the state of Florida to ascertain the votes cast for Andrea Gillum (Robles, 2018). In the 2017 assembly seat elections in India, in the state of Rajasthan the Electronic Voting Machines malfunctioned. Votes were allocated to the wrong candidate by the voting machine (Chatterjee, 2017). These incidents all points to the mistrust, by a central stakeholder, in the form of e-voting used for the election process. It is important to note that mistrust in an e-voting system is not without consequences to the adoption of e-voting. The consequences has been that of outright rejection of any form of e-voting as is in the case of Netherlands (Cerulus, 2017); the limited national adoption of e-voting as is in the case of Canada (Elections Canada, 2018) and the hybrid adoption of manual and e-voting as in the case of the United states (Ballotopedia, 2019). If there were a great deal of trust in the e-voting system to facilitate a free and fair election, these consequences would not exist. Furthermore, these consequences are neither encouraging nor inspirational to countries that are yet to adopt e-voting. However, e-voting provides voting convenience for the voter, reduces the cost of facilitating the election process and facilitates the efficiency in the electoral process (Anwar, 2009). However, despite these advantages, because of the inability of e-voting to facilitate trust, most countries in the world do not embrace e-voting (NDI, 2013) (evoting.cc, 2015). Therefore, there is the need to facilitate trust. If humans cannot trust themselves because the stake of the elections are too high, then technology designed to facilitate trust could be the arbiter. In this case, blockchain.

Therefore, to develop the proof of concept, the question answered in this chapter is "how can blockchain be utilized to facilitate trust in an election process?" In order to answer this question, the election process in Ghana is used as a test case. Ghana was chosen as a test case because the trust framework designed for the national elections can be transposed into the blockchain. Ghana from 1992 has emerged as a relatively stable and peaceful democracy. The country has experienced peaceful transition of power, back and forth between incumbent parties and opposition parties. However, despite these laudable credentials, the election process has not been without issues of mistrust between the central stakeholders. Though most of these challenges are handled peacefully one of them had to be settled at the Supreme Court. This was the litigation in the presidential election results of 2012 between the then incumbent President Mahama and the then challenger Mr. Akuffo-Addo, now President Akuffo-Addo. The then Mr. Akuffo-Addo lost the case at the Supreme Court. This implies that more has to be done to facilitate trust in the election process. Hence, the proof of concept in this chapter outlines how trust in the election process can be strengthened using blockchain.

This chapter is of practical significance to the facilitation of election processes in developing and developed countries as well. The scenarios presented are obviously inspirational but also practical. It can be modified to fit into any national scenario. This chapter is divided into 6 sections. The first section is the introduction. The second section provides an overview on the relationship between electronic voting, technology and trust. The third section provides an overview of the potential relationship between blockchain and the election process. The fourth section provides an insight into the proof of concept, using Ghana as a test case. These sections are followed by the discussions and the conclusion of the chapter.

5.2 Electronic Voting, Technology and Trust

An election is a formal process for selecting a person or a group of persons for either a public office or another position of responsibility by a person or a group of persons. Elections are evident in the creation of different forms of governments such as democracies, isocracies, autocracies, anocracies, totalitarianism and in some cases monarchies and other forms of governments. However, the election process for each type of government are very dynamic and the election process varies on the bases of the mode of election, process of election, cultural influence and institutions governing the electoral process. The level of participation of the electorate in these elections in the different forms of government also vary as well. In some cases, participation of the electorate in the election process may be limited. In other cases, a closed group of privileged persons could be mandated to elect a person to a political office. What is common, though among the various forms of governments, is that an election can only occur if there are more than one candidate to choose.

However, the form of government of interest in this chapter is democracy. This is because almost six in every ten countries in the world is a democracy (Desilver, 2017). The common denominator in a democracy is that the electorate choose the elected leader. Often they vote for the elected leader. They could vote for the leader directly or indirectly by electing persons who will elect their preferred leader. Although the voting process in different democracies vary, the elements of the voting process seem to be constant. How these elements are operationalized varies. The elements include the (1) registration of voters, (2) verification of voters, (3) casting of the ballot or voting, (4) collation and transmission of results and (5) the declaration of the eventual winner(s). These elements are governed by a set of rules defined by electoral agency in each jurisdiction. The electoral agency is often guided by the constitution of the jurisdiction. The constitution and the rules of the electoral body forms the common understanding by which the electoral agency, electorate, the political parties and the contesting candidate abide by. One could call it the basis of "fair play" in the election process. It also forms the basis of trust between these stakeholders.

However, this trust can be broken when a particular stakeholder feels that one of the other stakeholders is operating outside the laid down rules or constitutional boundaries governing elections. Examples of such breach in trust can be found in developing countries such as Malawi, Bangladesh, Kenya etc. (Md & Rawnak, 2018; Roberts, 2012; Taylor, 2018). Such a breach in trust could result in electoral litigations, demonstrations and either pre or post-electoral violence. Examples include the Philippine elections of 2018, the 2007 Nigerian election and the 2007 Kenyan elections to mention a few (Al Jazeera News, 2018; Dercon & Gutiérrez-Romero, 2012; Flores et al., 2015).

In order to curb these excesses and reintroduce trust into the process, different countries and jurisdictions have adopted different forms of e-voting technology. The idea here is that technology will limit human intervention in the process, thereby resulting in an election process facilitated in an atmosphere of trust. The processes adopted by different countries are illustrated in the following categories.

- **Voters' registration and verification**: Most western countries already have a well-established electronic national register, linked to national IDs. This leads to the facilitation of trust by the stakeholders in the integrity of the potential voter's register. However, in most developing countries in Africa, various card and of recent Biometric Verification Devices (BVD) technologies have been adopted. These efforts are made to certify that the person, whose name is on the voter's register, is the person coming to vote. Hence facilitating trust in the integrity of the voters register.

- **Balloting**: Globally, trust between the stakeholders is also elicited by the adoption of different technologies for the balloting process. This trend is evident in developed and in some developing countries, even though they are very few. In Africa, Namibia and Congo seem to be at the forefront in the adoption of e-voting. In Namibia, the Electronic Voting Machines (EVM) and Direct Recording Electronics (DRE) are adopted for the balloting process in national elections (WFD, 2018). DR Congo, adopted EVMs for the 2018 general elections (Ross & Lewis, 2018). However, they were unable to utilize the EVMs nationwide due to the destruction of about 8000 EVMs at the Electoral Commission's warehouse in Kinshasa (Mulegwa & Treasury, 2018). In Asia, India has been experimenting with Internet voting in the state of Gujurat (Singh et al., 2017). In Malaysia, political parties had adopted e-voting for balloting purposes while the Thailand Electoral Commission did consider adopting e-voting for the up-coming national elections which has been post-poned from 2017 (see (Mongkolnchaiarunya, 2016) (Strait Times, 2018) . Similarly, Khazakstan implemented e-voting but later rejected it (Gibson, et al., 2016). In South America, Brazil and Venezuela use the Direct-Recording Electronic (DRE) machines for their general elections (WFD, 2018). In developed countries, Estonia and Switzerland are at the fore-front of the adoption of e-voting. Both countries actually use Internet voting for their national elections (Kotkas, 2014). In the United States a range of e-voting technologies, ranging from e-mail, fax, DRE, Punch card etc. are used to facilitate elections in different part of the country (Ballotopedia, 2019).

Although e-voting technologies have been used to facilitate trust, these technologies does not have sufficient capacity to facilitate trust. The mistrust could be due to:

- **Human intervention**: The re-introduction of human intervention, this time in a negative manner, via hacking of the e-voting systems creates a breach in trust. This challenge is greater in the collation and transmission of results. In the cases where EVMs and DREs are used, the collation of the result are performed by the manual tabulation of the total results recorded and stored in different EVMs or DREs. In certain cases, these results can be manipulated (ITU Denmark, 2017).

- **Machine malfunction**: In some elections, DREs have been reported to malfunction. An interesting case is the 2012 Virginia elections where the DRE malfunctioned (Cho & Rein, 2003). However, it is important to note that in the case of the DRE, the total votes recorded in the machine could be transmitted online to the election centre. This has also been the case with EVMs as mentioned earlier in the case of India (Chatterjee, 2017).

In a sense, one could say that current e-voting technologies have failed in the facilitation of trust in the electoral process. This is one of the reasons why more than three quarters of countries in the world have not adopted e-voting technologies (evoting.cc, 2015; NDI, 2013). Although e-voting technologies have failed to facilitate trust in an electoral process, this does not imply that technologies cannot be used to facilitate trust. This is where Blockchain has an edge. Blockchain is designed with the trust by default principle and it has been used to facilitate trust in test elections. This is why blockchain is proposed in this chapter as a tool to facilitating trust in the electoral process.

5.3 Blockchain and Election Process

The potential of blockchain towards facilitating trust in an election process is inherent in its properties. This section provides an overview on blockchain, its properties and how these properties can facilitate trust in the election process.

5.3.1 Overview of Blockchain Technology

Blockchain is a technology that enables the movement of digital assets from one entity to another. A blockchain is a secured distributed ledger technology consisting of a growing list of transactions or records (Rennock et al., 2018). The blockchain network is decentralized peer-to-peer network. The peer or nodes create and validate transactions or smart contracts that occur between nodes. Each validated transaction is called a block. This block is added to a previous block forming a chain of records. Hence the name blockchain. Once the nodes validate the transaction, the data in the smart contract cannot be changed without validation. All the nodes in the network or a group of delegated nodes can make the validation. Hence, blockchain facilitates consensus.

Blocks in the blockchain are secured with a hash key. The hash key identifies the block and all of its contents (data) and this hash key is unique to each block. When a new block is added to a blockchain, the new block possesses its own hash key and the hash key of the previous block (ibid). Hence, the new block points to the previous block, forming a chain of secured completed time stamped digital transactions. If one of the blocks is tampered with, the hash key for the following block changes. Such a process will invalidate the subsequent blocks because they do not have the valid key of the previous block. To avoid such tampering, the proof of work concept or proof of stake is deployed (ibid). This is an algorithm used to validate the block before it is added to the blockchain. The proof of work process slows down the creation of a block, making it challenging and cumbersome to change and recalculate the hash key of all the blocks to make the blocks in the blockchain valid again. The proof of work concept ensures that data stored in the blockchain are immutable

The decentralized peer-to-peer network topology of blockchains also facilitates data security in blockchains (BlockchainHub, 2018). This is because other members of the peer-to-peer network possess the same blockchain. Whenever new blocks are added to the blockchain, the blockchain in every node is modified. The dynamics of the node synchronization process in blockchains can be explained using the principles of the ledger development and delivery processes. These principles are the Open Ledger principles, the distributed ledger principle and the miner principle. (Rennock et al., 2018; Zyskind et al., 2015).

- **Open ledger**: In an open ledger, there is a centralized blockchain visible and operated upon by every node in the network. Every validated transaction conducted between the nodes is added to this ledger. Hence, transactions performed in the open ledger are real-time and transparent to all the nodes. If digital assets or tokens are to be transferred from a smart wallet of one stakeholder or node to another within the network. Every stakeholder, including the receiver of the transaction, in the network is notified of the transaction. Furthermore, every stakeholder can verify that the initiator of the transaction has the needed digital assets to initiate the transaction. If the initiator does not have the needed digital assets, the initiator is ignored and the transaction will neither be validated nor received. But if the receiver can verify that the initiator has the digital assets to perform the transaction, if the receiver intends to honour the transaction, the transaction will be accepted and left for the relevant nodes to validate the transaction. Generally, in an open ledger, the nodes can also see and trace who initiated the transaction, when the block was created and when it was added to the ledger. Hence, the provenance of each transaction can be ascertained. This makes the blockchain a source of truth for the transactions.

- **Distributed ledger:** In the distributed ledger, every node has the current copy of the blockchain. Once a new block is created, the copy of the blockchain on every node is updated. This makes it impossible for one to modify a blockchain because the person has to tamper with the blockchain of all the nodes in order to make any changes. Such an attempt is quite daunting due to the tedious proof of work involved. Furthermore, when a new stakeholder or node is added to the network, a full copy of the existing blockchain is added to that node.

- **Miners:** Miners exist in a distributed network and in the cryptocurrency setup. They are delegated nodes with the mandate to validate transactions to be added to the blockchain. When an intent to perform a transaction between parties have to be announced, miners often compete on who will be the first to validate the transaction. The proof of work involves the identification of the key of the previous block through some complex computations. This will enable the miner to create a new block and add it to the existing blockchain and distribute it to the nodes. There is an incentive for the successful miner. In the case of cryptocurrency, the incentive is financial. The validation process is aimed at verifying the existence of the digital assets announced by the announcing party.

These principles provide additional security to the blockchain as well as facilitate trust. Hence, blockchain as a technology is designed based on the concept of "trust-by-default". The trust is facilitated technically using cryptography and socially by enabling of transparency and consensus at the nodes. The cryptography in blockchain facilitates immutability and finality in the blockchain, while the social dimensions are exhibited via consensus and provenance.

There are three types of blockchain. These are the public, private and federated blockchains. Public blockchains are open to all and permissionless (BlockChainHub, 2018). Federated blockchains is not open to the public; it is permissioned and operated by either a group or consortium (ibid). Consensus on such a blockchain is only limited to the consortium or group. Other members of the platform are able to perform transactions but not validate them. Either individuals or groups own private blockchains and it is permissioned (ibid). It is not open to the public and the nodes are accepted by invitation of the owner of the blockchain. The rules governing consensus is decided by the owner of the platform. There are obvious potentials for blockchains in the facilitation of an election process. This is discussed in the next subsection.

5.3.2 Potential for Blockchains and Elections

In the previous section, certain properties of blockchain was discussed. These properties were consensus, transparency, immutability and finality (Cosset, 2018). These properties coincidentally are some of the hallmarks of an election process. This section provides an insight into the similarities between some of the hallmarks of an election process and the properties of blockchain.

- **Consensus:** Consensus is very important in an election process. The three main stakeholders have to agree on the rules, processes and outcome of the elections. Such rules are often enshrined in the national constitution. With respect to the process, there is need for consensus on the integrity of the voter's identity in the voters' register; consensus on the ballots cast; consensus on the results collated and consensus on the result announced. Where there is a lack of consensus, certain parties are bound to become aggrieved. Examples of such grievances can be identified around the globe. In the 2018 US mid-term elections, there were minor agitations on the lack of consensus on the legitimate means of identity needed to register to vote in some states. In African countries such as Nigeria, Mali, Ethiopia and Ghana, there has been a lack of consensus on the integrity of the voter's register (UNECA, 2005). In the 2018 Brazilian general elections, there lack of consensus on the eligibility of voters as found in the voters register (Lloyd, 2018). This led to the removal of names of certain voters from the voter register. The agitation led to a litigation. The supreme electoral court had to mandate persons whose names were removed from the voters register to vote. In the Pakistani elections, there was lack of consensus among the stakeholders on how to secure the voting process (Shakil, 2018). The army was employed to monitor the collation of the ballots. In the just concluded 2018 elections in DR Congo, there was a lack of consensus on who won the elections. The consequence was that some parties refused to accept the result of the elections. Hence, consensus is an important facet of an election process. As described earlier, blockchain thrives on consensus. Therefore, it has the potential to facilitate consensus on the various process of an election process.

- **Provenance:** In an election process, the transparency of the process from the voter registration/verification process to the announcement of the results is critical. The electorate require transparency in the manner their ballot is handled. Furthermore, they want to be sure that their votes are registered to their choice or candidate and that the vote counts in the outcome of the elections. The political parties and candidates, require that only eligible persons are able to vote; that they vote once; and that their vote counts. The Electoral Commission is interested in making sure that the inherent demands of the electorate, political parties and the candidate are catered for in a transparent manner during the national election process. In most cases, this is not so. This is evident in the case of litigations. The audit trails of the process are often cumbersome and time consuming. This is because of the volume of data and documents involved. This is also the case in the case of Internet voting (i-voting), which is an e-voting process where the electorate vote over the Internet. However, with blockchain every node can trace the history of the transactions made because they have a copy of the blockchain. It is also easy to generate in the case of a litigation. Generally, blockchains in an ideal situation should limit litigation. This is because the information or data recorded concerning the votes are accurate and traceable.

- **Immutability:** One of the challenges in an election is the tampering of either the voters registration, ballots cast, the collation and transmission of results and the results. In a blockchain, as explained earlier, it is difficult to alter a block in the blockchain. Hence, data stored in the block are immutable and trustworthy.

- **Finality:** A blockchain possesses only one ledger for the whole network (Cosset, 2018). The ledger is transparent as explained earlier. Therefore, in an election, every stakeholder serving as a node in the blockchain would have validated every block before it is added to the ledger. Hence, the ledger is the source of fact. In such a scenario, there will be no need for litigations as every party to the blockchain would have certified either the voters register, the ballots cast, or the results collated.

Therefore, blockchains has the potential of providing benefit to an electoral process. This is why it is promoted in this chapter.

5.4 The Potential Use Case – The Ghanaian General Elections

5.4.1 The Ghanaian Election Process

The Ghanaian general elections consist of the presidential and the parliamentary elections. The election process is as follows:

- Pre-election Process:
    1. **Registration of voters:** New voters, who are 18 years and above are registered into the Biometric Register (BR) using a BR kit. The kit includes a Laptop, a figure print scanner, hand scanner, and a printer.

    2. **Exhibition of the voter register**: To ascertain the integrity of the voter register, voters are given the opportunity to verify their registration details. This is done as an extension of the voter registration.

    3. **Transfer of votes**: During this stage, citizens are allowed to transfer their votes from one region to another. This possibility is given to those who have relocated. Transfers within districts are not permitted. These modalities concerning transfers are decided by IPAC (Inter-Party Advisory Committee). After these activities, the application for transfers of votes and proxy voting, followed by a

notice of election where the general public is informed about the upcoming elections.

- Election process
    1. **Voting process**: On Election Day, voting officially starts from 07:00 am and ends at 17:00, and each voter is required to be at a designated polling station to cast his/her vote.

    2. **Verification of voters:** When it gets to the turn of a voter, he/she has to be verified in order to be certain if the voter is eligible to vote at the polling station. At this stage, the first point of contact is the person in charge of the Voter Reference List (VRL). The VRL contains the list of all registered voters details for a particular polling station. The details of the voter is checked against the VRL and if found the record will then be marked. Once the voter records are found in the voter's reference list, he/she will now have to go through the biometric verification using the Biometric Verification Device (BVD), and the voter's fingers are marked with indelible ink to indicate the voter has been verified.

    3. **The casting of the ballot or voting**: After the voter has been verified, the voter then proceeds to the presidential ballot issuer to be issued with the presidential ballot paper. The voter then goes to the presidential voting screen to thump print the ballot and cast the vote by dropping the ballot into the presidential ballot box. Next, the voter moves to the parliamentary ballot issuer to be issued with the parliamentary ballot paper and the voter then cast the parliamentary ballot.

- Post – Election (Collation and transmission of results)
    1. **Collation:** After the voting process has ended at a polling station, the polling officials perform ballot accounting. In the ballot accounting process, the total ballot issued out is checked against the verified voters by the BVD as well as the ticks or marked records in the voter reference list, and secondly, all the ballot have to be checked if they contain EC stamp.

    2. **Transmission:** After this pre-process activity, the ballots are sorted according to the candidates on the ballot paper and counted by the returning officer at the polling station. The results are then recorded on the declaration results sheet (also known as the pink sheet). The polling station results for both presidential and parliamentary

are announced at that polling station. The results recorded on the declaration results sheet at the various polling station within a constituency are then send to the constituency collation center via physical transportation to the collation center or by fax. At this point, transmission of the collated results has to be transmitted to the national Electoral Commission headquarters, and it can be done in two ways depending of the mode of transmission (electronic transmission or manual transmission by fax). For manual transmission the collated results from the polling stations at the constituency collation center will be sent to the district EC office, the district EC office will fax the results to the regional EC office and the regional EC office will then fax it to the national EC headquarters. For electronic transmission the constituency result obtained from the various polling stations are captured in an electronic form (using a software application) at the constituency and it is transmitted electronically to the national EC headquarters.

- The declaration of the eventual winner: Declaration of winners only happens at the National office for presidential and constituency centers for the parliamentary election. After all the constituency results have been received and verified, the eventual winner for the Presidential is declared. Parliamentary results are declared at the constituency collation centers.

5.4.2 Challenges in the Ghanaian Election Process

Despite the peaceful transition in government, elections in Ghana are not without issues. These issues pale in comparison to issues surrounding the electoral process of next-door neighbours such as Togo and Cote de Ivoire. However, there are challenges that could or might have damaged trust in the electoral process in Ghana. These challenges are evident in at the voter registration and verification, balloting and the collation and transmission of results. The breakdown of the challenges are as follows:

- **Lack of trust in the voter registration and verification challenges**: In 2012, Ghana introduced the biometric voter verification process (Effah & Debrah, 2018). This initiative was to rid the bloated voter register of underage voters. Though the introduction of the biometric voters registration was successful at certain polling stations, some of them were faulty or could not verify the fingers of certain individuals (ibid). Obviously, this disenfranchised some voters and the results for the presidential elections in that national elections were contested in court. Furthermore, despite the adoption of the biometric voter verification, the credibility of the voter register is still called to question (Code of Ghana, 2017; EU EOM, 2016). Hence, there is a lingering issue of trust, when it comes to voter registration and verification in Ghana.

- **Lack of trust in the balloting process:** In the 2012 elections, the Coalition of Domestic Elections Observers (CODEO) identified violation of voting procedure, intimidation of voters, vote buying and abrupt suspension of voting as some of the challenges in the balloting process (Code of Ghana, 2016). In the 2016 elections, political parties as a way of protecting their interests at the polls had vigilante groups at polling stations (Code of Ghana, 2017). These problems has the potential of suppressing the fundamental human right of a voter towards voting for a candidate of his or her choice. Hence resulting in lack of trust in the voting process by the electorate and minor political parties.

- **Collation and transmission of results:** Before the 2016 elections in Ghana, election results were collated and verified at the polling center, transferred physically to the regional electoral office for further fax or physical transmission to the Electoral Commission (Commonwealth, 2012). In 2016, results were scanned from the polling stations and transmitted electronically to the Electoral Commission (EC, 2016). The challenge as identified by the CODEO was that the political parties had a read and write privilege to the transmitted results (Code of Ghana, 2017). This does not imply that the results were tampered with. However, it is a potential challenge that could result in a bridge of trust in the election result.

Based on these challenges, it is evident that the Ghanaian electoral system is in need of tools to facilitate greater trust in the electoral system. How this could be achieved with blockchain is described in the next section.

5.5 Proof of Concept

5.5.1 Current Trust Framework

The Electoral Commission conducts elections in Ghana, as described in the previous section. The relevant stakeholders are the Electoral Commission, political parties and the electorate. These parties have to trust themselves in order to facilitate free and fair elections. Currently in Ghana, trust is built in the following ways, namely:

- **Open ballot counting at polling station:** The open counting of ballots occurs at the polling stations by the Electoral Commission official. The counting is performed in the presence of party agents and the electorate present at the polling station.

- **Open verification of results by political party and EC officials at polling station:** once none of the stakeholders contests the outcome of the result, then the representatives of the EC and party agents sign it.

- Verification of the comprehensive results by the political parties and EC officials at the EC's national office.

Based on this short summary, it is clear that the Electoral Commission drives the process, while the political parties and the electorate provide quality assurance for the process. This trust process will be transposed into the blockchain. However, the electoral process itself will not be transposed into the blockchain.

5.5.2 Potential Trust Framework with Blockchain

To facilitate a general election in Ghana, two blockchains will be needed. One for the presidential elections and one for the parliamentary elections. These blockchains will be federated blockchains (scenarios 1 & 2) and private blockchain (scenario 3), developed, owned and operated by the Electoral Commission of Ghana. The blockchain will be connected to their existing traditional databases for the tabulation and computation of results. If the Electoral Commission adopts i-Voting in future, then the blockchain will be linked to the i-voting system.

Depending on the scenario, as will be described later, the electorate could be active nodes in the blockchains. This would have been ideal if every voter in Ghana had access to a mobile or desktop device in addition to Internet connectivity. However, since that is not the case, there is also the option of a polling station being represented as a node in the blockchain as a robot user. As a way of transposing the current trust framework to blockchain, Consensus in the network is delegated to the Electoral Commission and the political parties as mentioned in the Table 5.1.

In each blockchain, transaction only occurs between the electorate and the Electoral Commission. In the situation where no electorates are connected to the blockchains, the transaction will be between the polling station (robot accounts) and the Electoral Commission. The account of each electorate connected to the blockchain will possess a smart wallet. To ensure the anonymity of the voter, only the voter's registration number will be made visible on the

Table 5.1 Potential nodes for the federated Blockchain

No	Item	Potential nodes
1	Owner of Blockchain networks	Electoral Commission of Ghana
2	Permanent nodes	Electoral Commission of Ghana
		Political parties
3	Optional nodes	Electorate (voters)
		Polling station (robot accounts)
4	Delegated nodes to validate transaction	Electoral Commission of Ghana
		Political parties
5	Delegated nodes that perform	Electoral Commission of Ghana
	transactions	Electorate (voters)
		Polling station (robot accounts)

blockchain account. The tokens in the smart wallet will be two e-ballots, one for each election. In the case of polling station (robot accounts), the number of e-ballots will match the number of voters registered in that polling station.

5.6 Conceptual Scenarios

Three scenarios are proposed. The Electoral Commission of Ghana in their bid to conduct national elections using blockchain can adopt these scenarios. The first scenario involves the facilitation of remote voting with the two blockchains. This proposal is an end-to-end system. The second scenario involves the facilitation of indirect voting on the blockchain from polling booths. The third scenario is on how to incorporate blockchain to facilitate vote storage in the current Ghanaian election process.

5.6.1 Scenario 1 (Remote Voting)

The first would be an end-to-end system. The class of stakeholders in the system will include the eligible voters, the party representatives and the electoral officials from the Electoral Commission. This is represented in Figure 5.1 below. The various subsets of these class of stakeholders will serve as nodes in the blockchain network. The prerequisite for this scenario is the adoption of I-voting in Ghana; the access to broadband Internet access by every citizen of voting age; and the access to unique mobile and desktop applications by every Ghanaian of voting age.

In the infrastructure architecture, the blockchain Network will be at the edge of the infrastructure of the Electoral Commission as seen in the Figure 5.1. In this network, every node will have a copy of the blockchain in their account. However, the voters will have a smart wallet containing

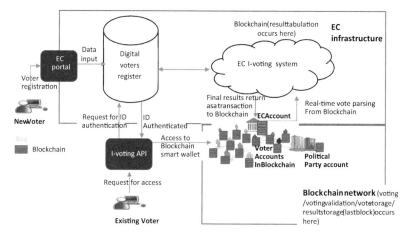

Figure 5.1 End-to-end Blockchain solution.
Source: Williams & Agbesi (2019).

the ballot in their account. These voters will vote via a smart contract transaction with the Electoral Commission. However, these contracts will not become blocks of completed votes until the political parties and the Electoral Commission have validated it. The reason political parties have to validate every vote is to facilitate trust in the voting process. Else, the alternative is that these parties validate the final result. This will be explained in the scenario. The core Electoral Commission infrastructure will be the Digital Votes Register (DVR) and their i-Voting system as seen in Figure 5.1. The DVR will enable the registration of new voters and the authentication of existing voters. Authenticated voters can access the blockchain account. The i-voting system consists of the i-voting client – server database architecture. The only difference is that data from the client-side is now channelled through the blockchain to the server as seen in Figure 5.1. In the server is a database that retrieves real-time data from the blockchain. The computation of results will occur in this database. Once the election is over and the results computed, the EC will then feed the result into the blockchain. Here the political parties will validate the results, a block will be created and it will be added to the blockchain. Here is how it works in a scenario.

- **Scenario construction**

Step 1

Citizen Kofi, a registered voter, receives an SMS notification to participate in the national elections. He follows the following steps.

Citizen Kofi

Election commission Blockchain portal

Figure 5.2   Access to both Blockchains on the Electoral Commission's portal.
Source: Williams & Agbesi (2019).

1. He accesses the national electoral i-voting platform using his electoral identity on a mobile phone application of desktop application.
2. The system ask for his digital signature or biometric authentication.
3. Once verified, he is redirected to the Electoral Commission's blockchain portal. Here he can choose which elections he intends to be a part of as seen in Figure 5.2.
4. He is granted access to his account on the Electoral Commission's blockchain. He has access only to one non-renewable token, which enables him to vote once.
5. He selects the candidate of his choice and clicks on the "vote" button. The program allows him to choose only one candidate. Once he has voted, the action is irreversible.
6. The political parties and the Electoral Commission officials on the blockchain, receive a notification on the blockchain that a vote has been cast. They can only see the transaction from Kofi's voters' registration number and the selection made by Kofi. No other data owned by Kofi is transmitted.
7. The relevant political party and Electoral Commission officials click on the "validate button". An automated "proof of work" is performed in the blockchain on the transaction.
8. The validated vote is added as a block to the blockchain. It is further distributed to all the nodes.
9. In real-time, data from that block is parsed via xml to the Electoral Commission's traditional database.
10. He receives a confirmation by SMS that his vote has been cast.
11. His vote is mapped, in the traditional database, to his voter's registration number which is linked to the Digital Voter's register. In this case the voter's registration number is the primary key.
12. He then exits the first election and repeats the same process for the next elections.

Step 2

Once the voting process ends, result computation is made on the Electoral Commission's database. The overall results are fed into the blockchain as a transaction by the Electoral Commission. The political parties validate the transaction and the block is added to the blockchain. At the end of both elections. The transactions will be visible to all parties.

5.6.2 Scenario 2 (Polling Booth Vote)

The second solution is not an end-to-end system but it is similar to the first solution. However, there are minor differences. The first difference is that the electorates are not the nodes. They are replaced by pooling stations. Each polling station is a node on the blockchain network. The Electoral Commission and political party are still nodes in the blockchain network. The second difference is that the voting technology is not necessary i-voting. It could also be a DRE machine. The important thing is that the e-voting technology used should be connected to the Internet and results should be transmitted directly to the blockchain. At the blockchain, the smart wallet of each polling station node is equipped with the number of tokens that match the number of voters at the polling station. Unlike scenario one, the voter verification process is performed separately. Currently, biometric voter verification process is used in Ghana. This feature is maintained in this scenario as seen in Figure 5.3.

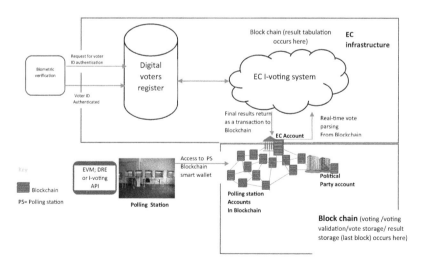

Figure 5.3   Blockchain architecture interfaced with DRE or I-voting.
Source: Williams & Agbesi (2019).

The minor difference is that, in this scenario, data from the DVR are used to verify the biometric information in the biometric verification machine.

- **Scenario construction**

Step 1

Citizen Alice visits the polling station to vote. She arrives at her polling station. She will follow the following procedure.

1. She undergoes biometric verification and it is registered in the digital register that she came to vote.
2. She is then directed to the polling booth equipped either with an i-voting or a DRE terminal.
3. Just as in the case of citizen Kofi, she is presented with the possibility of voting on one election before the other. She selects the first election.
4. She is presented with a list of candidates on the screen. If it is a touch screen she selects her candidate by touching the picture of the candidate. If it is not a touch screen she used the input device to select her candidate. If she is not computer literate, the polling agent will assist her.
5. Her choice is transmitted in real time over the Internet using xml to the polling station's blockchain account.
6. On the blockchain, data from the xml transmission is displayed on an e-ballot in the account's smart wallet.
7. The ballot will only reflect the voter's electoral number and the candidate voted for.
8. The polling agent sends her to another polling agent who will grant her access to the polling station's blockchain account. Here she will verify that the details on the e-ballot reflects her initial choice.
9. If the answer is yes, the polling agent will then commit the vote for validation by the Electoral Commission and the political parties.
10. If the answer is no, she has to vote again.
11. Once the vote is validated. The data is retrieved from the blockchain in real-time to the Electoral Commission's traditional database housing the results. Here the vote is matched to the voters electoral number, just as in scenario one.
12. She receives an acknowledgement by the polling agent of a successful vote cast.
13. Her vote is mapped, in the traditional database, to her voter's registration number which is linked to the Digital Voter's register. In this case, just as in scenario one, the voter's registration number is the primary key.

14. She then exits the first election and repeats the same process for the next elections.

Step 2

This step is the same as in scenario one.

5.6.3 Scenario 3 (Result Storage Only)

This is also not an end-to-end system. This scenario incorporates the current election process in Ghana and the blockchain is used at the end of the day to store the outcome of the certified votes cast. These are polling station votes, constituency votes, regional votes and the national votes. This will be a private blockchain owned by the Electoral Commission. The other nodes will be political party nodes. The reason for the restriction on the class of nodes is to extend the offline result verification exercises online. Results sent to the Electoral Commission as mentioned earlier are sent electronically, either faxed or transmitted physically. At every station, party officials are allowed to verify that the figures presented by the Electoral Commission are valid. This they do by receiving information from their agents at the polling stations and at the constituencies. The introduction of the blockchain would provide a psychological boost towards trusting the process. This is because all validated votes sent to the blockchain are immutable.

When the results of each stages of the votes are fed into the blockchain by the Electoral Commission, the political parties can validate the results again, this time on blockchain. If this process is performed before the results are announced, it will facilitate trust between the political parties and the EC in the election process.

5.7 Discussion

Though these scenarios are proposed, the fact is that the third scenario is the only feasible solution that could be deployed in Ghana now. This is because it is a less costly solution, it requires less computing power. However, it facilitates trust in only one of the election processes mentioned earlier. Nevertheless, the best solution for facilitating trust in an election using blockchain would be scenario one, followed by scenario two and then three. Scenario one will facilitate greater transparency and trust between the major stakeholders. Voters would be able to vote at their convenience and there is very little room for party vigilantes to intimate the voters. It might be the most expensive solution and may require a lot of computing power. Nevertheless, it is less costly than the undesired consequences that could occur if an election process goes wrong. Scenario two will facilitate transparency and trust to a certain degree, but there is the possibility of the vote tampering between the polling booth terminal and the blockchain. If the data transmission via the Internet is intercepted, votes could be manipulated. This is why an extra measure is added, where the polling agent has to verify that the transaction generated from the polling booth is what reflects in the blockchain. This would require having at least two polling agents. One polling agent will aid the voter in the casting of the ballot. This extra measure will ensure offline transparency for the voter.

Although Ghana is not ready for scenario one and two, it is not far away from achieving these scenarios. To achieve scenario one, there is the need for to develop the requisite digital infrastructure, promote civic engagement and facilitate a change management process between the major stakeholders as outlined in the Table 5.2 below.

The Electoral Commission already has a voter register that should be digitalized. They also have the biometric means of verifying voters, which should be linked to the unique number in the voter ID. They also have a database, which could receive data from the blockchain. They need an i-voting client-server architecture which would incorporate the digital register, biometric verification and the results database to facilitate scenario one. In addition, the electoral laws have to be modified to fit this process. The Electoral Commission could develop such an infrastructure with funding from donor agencies

Table 5.2 Prerequisites for scenario 1

No	Electoral commission needs	Electorate needs	Political parties
1	There has to be a digital national register	Every voter should have a unique digital ID and signature	Political parties have to agree on adopting blockchain.
2	The EC must have an existing I-voting infrastructure	Every voter must have an access device	The political parties have to upgrade their IT infrastructure
3	Electoral laws must be modified to fit the process	Digital illiteracy of voters should be eradicated Every voter must have access to the Internet Need for devices that are friendly to the disabled	The relevant party agents should be digitally literate

and the national parliament. Obviously, this would require an upgrade in their technical competence to manage such an infrastructure.

A sizable portion of the population are not digitally literate. However, digital literacy in itself is not the reason people do not adopt technology (Williams et al., 2016). However, in the case of e-voting, it could lead to the disenfranchisement of a voter. Therefore, the Electoral Commission has to facilitate knowledge management initiatives that would support the e-voting process, preferably in the local dialect. In addition, scenario one will not be possible if eligible voters do not have access to the Internet and the relevant access equipments. Such equipment should be friendly to the disabled. To make this happen, national government has to ensure that everyone has access to the Internet. Currently there are cost effective ways of deploying Internet in rural areas (Nungu & Pehrson, 2011). These possibilities even extend to 5G (Williams, 2018). Furthermore, national governments should promote initiatives that will facilitate the diffusion of access equipments to rural and disabled people. This is where the government of Ghana could partner with NGOs and international Development Agencies. Lessons could be learnt from the Bill and Melinda gates foundation initiative in Vietnam (Do Manh et al., 2015).

The political parties are very crucial to the adoption of scenario one. In Ghana, there is a body called the Inter-Party Advisory Committee (IPAC). It is a stakeholder consultative forum organized by the Electoral Commission to consult with the political parties. This forum is a good tool towards facilitating change management in the electoral process. Hence carrying this body along in a transparent manner in the deployment of scenario one would enable the political parties to accept the policy, upgrade their systems to become nodes in the blockchain network.

The requirements for scenario one also applies for scenarios two and three but with minor difference to scenario one. These differences apply only to the Electoral Commission and the electorate. The electorate do not need Internet access or mobile applications for scenarios two and three. They only need to go to the polling station with their voters ID. Hence, scenarios two and three takes into consideration income inequality as well as Internet access inequality between persons who are poor and those living in rural areas. However, for scenario two, the Electoral Commission has to replace the ballot box with an EVM or most appropriately a DRE in addition to the blockchain system. So one could say that the cost of developing and operating each scenario decreases from scenario one to three.

The problem in the proof of concept presented in this chapter is that of anonymity. The anonymity of the voter and that of the vote cast. In this chapter, the voter's registration number is used to reduce the visibility of the identity of the voter. In a blockchain anonymizing the user account is possible. However, in the case of Ghana's election being able to sort the votes in the Electoral Commission's database into polling station, constituencies and regions matter. Hence, the transmission of the user voter's registration number will enable the Electoral Commission's database to sort the votes. Furthermore, it is assumed in this proof of concept that the flux of votes needing validation will not give the validators' time to perform a background check on the user and whom he or she voted for.

The proof of concept presented in this chapter does not apply to Ghana alone. The process flow is Ghana centric, but concept itself is not. It can be applied anywhere but with modification of course. What one should consider when designing such a process should be the modification of the election process towards the design of blockchains and not vice versa. This might require updating existing electoral laws, which will in turn enable the election process to accommodate blockchain. Such a system should be an end-to-end system as much as possible. However, the readiness of the voters should not be overlooked. This includes access readiness and "usage readiness". Access readiness involves the voter having access to the Internet and the relevant access equipment. "usage readiness" implies that they can use and are mentally ready to use the access device as well as the blockchain e-voting system.

Although the end-to-end solution is favoured in this chapter, the proof of concept has shown that there is flexibility in the deployment of blockchains in an election scenario. Hence, if the trust challenges occurs in one section of the election process, blockchain can be used to facilitate trust in that section. The important thing to note is how the blockchain will interface with the other processes. Finally, although many good things has been said about blockchain, in developing the proof of concept, it was realized that result tabulation is problematic using blockchain. However, the good news is that one can parse data from the block to an external database.

5.8 Conclusion

Blockchains may be regarded as a hype. This is because of how it evolved and what it has been used for. In many cases, blockchains have been confused with Bitcoin and other cryptocurrencies. However, if one decouples blockchain as a technology from the hyped service, it is possible to see different use cases for blockchain. These use cases are those that require trust and transparency between multiple parties. One of such use cases is the conducting of elections as illustrated above. As mentioned, there are trials on the use of blockchains for elections. In addition, elections in developing countries are tense moments. Apart from the trial by Agora in the sierra Leonean elections, blockchains have not been considered as a means of facilitating trust. In this chapter using the Ghanaian election process, a proof of concept of how blockchain could be utilized is described. Based on the scenario descriptions, one could say, bearing technical implementation difficulties that blockchains can facilitate trust in an electoral process. This chapter is aimed as serving as an inspiration for the use of blockchains in national elections and the scenarios aimed at providing initial requirement specifications to any agency that desires to implement such a system. This chapter concludes that it is possible to use blockchains in national elections.

References

AFP. (2014). "Namibian election first in Africa to use electronic voting machines," Available at: http://www.abc.net.au/news/2014-11-28/namibian-election-first-in-africa-to-use-electronic-voting/5927206.

Al Jazeera News. (2018). "Philippines: Election-related violence leaves 33 dead." Available at: https://www.aljazeera.com/news/2018/05/philippines-election-related-violence-leaves-33-dead-180514080913889.html.

Anwar, N. K. (2009). *Advantages and Disadvantages of e-Voting: The Estonian Experience*. Nanyang Technological University.

Ballotopedia. (2019). "Voting methods and equipment by state." Available at: https://ballotpedia.org/Voting_methods_and_equipment_by_state.

BlockChainHub. (2018). "Blockchains and Distributed Ledger Technologies." Available at: https://blockchainhub.net/blockchains-and-distributed-ledger-technologies-in-general/.

BlockChainHub. (2018). "What is block chain." Available at: https://blockchainhub.net/blockchain-intro/.

Cerulus, L. (2017). "Dutch go old school against Russian hacking." Available at: https://www.politico.eu/article/dutch-election-news-russian-hackers-netherlands/.

Chatterjee, A. (2017). "18 BJP Only EVMs Seized in Dholpur, Rajasthan," Available at: http://www.stateherald.com/2017/04/09/18-bjp-only-evms-seized-in-dholpur-rajasthan/.

Cho, D. and Rein, L. (2003). "Fairfax To Probe Voting Machines." Available at: http://www.washingtonpost.com/wp-dyn/articles/A54432-2003Nov17.html??noredirect=on.

Code of Ghana. (2016). "Ghana's 2012 Elections Irregularities (CODEO Report)." Available at: http://www.codeforghana.org/2016/04/19/irregularities-in-2012-general-elections.html.

Code of Ghana. (2017). "CODEO Communique." Available at: http://www.codeoghana.org/assets/downloadables/CODEO%20Communique_6April17.pdf.

Commonwealth. (2012). "Report of the Commonwealth Observer Group." Available at: http://aceproject.org/ero-en/regions/africa/GH/ghana-final-report-presidential-and-parliamentary-2/at_download/file.

Cosset, D. (2018). "The 4 characteristics of a blockchain." Available at: https://dev.to/damcosset/the-4-characteristics-of-a-blockchain-2c55.

Dercon, S. and Gutiérrez-Romero, R. (2012). "Triggers and Characteristics of the 2007 Kenyan Electoral Violence." *World Development*, 40 (4), 731–744.

Desilver, D. (2017). "Despite concerns about global democracy, nearly six-in-ten countries are now democratic," Pew Research Center, Washington, 2017.

Do Manh, T., Falch, M., and Williams, I. (2015). "The Role Of Stakeholders On Implementing Universal Services In Vietnam," in 26th European Regional Conference of the International Telecommunications Society (ITS), Madrid, Spain, 24-27 June 2015.

EC. (2016). "Clarification on Election Results Transmission and Contractural Process." Available at: http://www.ec.gov.gh/medias/press-release/107-clarification-on-election-results-   transmission-and-contractural-process.html.

Effah, J. and Debrah, E. (2018). "Biometric technology for voter identification: The experience in Ghana," The Information Society, vol. 34, no. 2, pp. 104–103.

Elections Canada. (2018). "A Comparative Assessment of Electronic Voting," Available at: http://www.elections.ca/content.aspx?section=res&dir=rec/tech/ivote/comp&document=municip&lang=e.

Eu Eom. (2016). "Eu EOM Ghana Presidential And Parliamentary Elections 2016: Final Report," 2016. Available at: https://eeas.europa.eu/sites/eeas/files/final_report_-_eu_eom_ghana_2016.pdf.

evoting.cc. (2015). "E-voting map." Available: https://www.e-voting.cc/en/it-elections/world-map/.

Flynn, P. (2017). "What Brexit should have taught us about voter mani pulation." Available at: https://www.theguardian.com/commentisfree/2017/apr/17/brexit-voter-manipulation-eu-referendum-social-media.

FMT. (2018). "PKR may go back to paper ballots after rowdy start to e-voting." Available at: https://www.freemalaysiatoday.com/category/nation/2018/09/23/pkr-may-go-back-to-paper-ballots-after-rowdy-start-to-e-voting/.

Gibson, J. P., Krimmer, R., Teague, V., and Pomares, J. (2016). "A review of e-voting: The past, present and future." annals of telecommunications – annales des télécommunications, vol. 71.

Huillet, M. (2018). "Russian Independent Electoral Watchdog to Pilot Blockchain for Voting System." Available at: https://cointelegraph.com/news/russian-independent-electoral-watchdog-to-pilot-blockchain-for-voting-system.

ITU Denmark. (2017). "USA: Virginia ditches DRE voting machines after ITU researcher's hack." Available at: https://en.itu.dk/about-itu/press/news-from-itu/2017/virginia-uk.

Kotkas, V. (2014). Security Analysis of Estonian I-voting System Using Attack Tree Methodologies, Tallin: Tallin University of Technology.

Linver, H. (2018). "Blockchain and Elections: The Japanese, Swiss and American Experience." Available at: https://cointelegraph.com/news/block chain-and-elections-the-japanese-swiss-and-american-experience.

Lloyd, R. (2018). "Brazil: the Supreme Electoral Court wiped out voters' reg-istrations." Available at: http://globalcit.eu/brazil-the-supreme-electoral-court-wiped-out-voters-registrations/.

Mongkolnchaiarunya, J. (2016). "A Serious Concern Over the First Use of E-Voting in Thailand." Available at: https://thediplomat.com/2016/10/a-serious-concern-over-the-first-use-of-e-voting-in-thailand/.

McConaghy, T., et al. (2016). "BigchainDB: A Scalable Blockchain Database," Aalto University, 2016.

Md, A. H. and Rawnak, J. (2018). "Parliamentary election and electoral violence in Bangladesh: the way forward," *International Journal of Law and Management*, 60(2), 741–756.

Mulegwa, P. and Treasury, K. (2018). "Presidential election in the DRC: Ceni postpones the elections to 30 December 2018." Available at: https://www.jeuneafrique.com/692324/politique/presidentielle-en-rdc-la-ceni-reporte-les-elections-au-30-decembre-2018/.

NDI. (2013). "Electronic Voting and Counting Around the World." Available at: https://www.ndi.org/e-voting-guide/electronic-voting-and-counting-around-the-world.

Nungu, A. and Pehrson, B. (2011). "Towards Sustainale Broadband Communication in Rural Areas," Lecture Notes of the Institute for Computer Sciences, Social Informatics and Telecommunications Engineering, vol. 63, no. 2011, pp. 168–175.

Pollock, D. (2018). "Who Created the Story of Sierra Leone's Blockchain Election?." Available at: https://cointelegraph.com/news/sierra-leones-fake-blockchain-election-hasnt-damaged-the-technologys-reputation.

Rennock, M. J., Cohn, A., and Butcher, J. R. (2018). "BlockChain Technology and Regulatory Investigations." Thomson Reuters.

Roberts, M. R. (2012). "Conflict analysis of the 2007 post election violence in Kenya," in Managing conflicts in Africa's democratic tradition, Rowman & Littlefield, pp. 141–154.

Robles, F. (2018). "Nearly 3,000 Votes Disappeared From Florida's Recount. That's Not Supposed to Happen." Available at: https://www.nytimes.com/2018/11/16/us/voting-machines-florida.html.

Ross, A. and Lewis, D. (2018). "In Congo, voting machines raise suspicions among president's foes," Available at: https://www.reuters.com/article/us-congo-election/in-congo-voting-machines-raise-suspicions -among-presidents-foes-idUSKCN1GL13W.

Shakil, F. M. (2018). "Imran Khan's PTI makes big gains in Punjab ahead of election." Available at: http://www.atimes.com/article/imran-khans-pti-makes-big-gains-in-punjab-ahead-of-election/.

Singh, V. P., Pasupuleti, H. and Babu, N. (2017). "Analysis of Internet Voting in India," in International Confernece on Innovation in Information, Embedded and Communication Systems (ICIIECS) 17–18 March 2017.

Strait Times. (2018). "Malaysia's PKR promises to have a more efficient e-voting system following glitches." Available at: https://www.straitstimes.com/asia/se-asia/malaysias-pkr-promises-to-have-a-more-efficient-e-voting-system-following-glitches.

Taylor, C. (2018). "Shared Security, Shared Elections: Best practices for the prevention of electoral violence," American Friends Service Committee, 2018.

UNECA. (2005). African governance report, Economic Commission for Africa, 2005.

Weintraub, M., Vargas, J. F., and Flores, T. E. (2015). "Vote choice and legacies of violence: evidence from the 2014 Colombian presidential elections." *Research and Politics*, 2(2), 1–8.

WFD. (2018). "Voting machine review." Available at: https://www.ceni.cd/assets/bundles/documents/voting-machine-review-wfd-ceni.pdf.

Williams, I. (2018). "Community Based networks and 5G Wifi," Ekonomiczne Problemy Uslug. vol. 131, pp. 321–334.

Williams, I. (2018). "Community Broadband networks and the Opportunity for e-Government Services," in Advanced Methodologies and Technologies in Government and Society, IGI, pp. 173–185.

Williams, I., Kwofie, B., and Sidii, F. (2016). "Public Demand Aggregation as a means of Bridging the ICT Gender Divide." in Overcoming Gender Inequalities through Technology Integration., IGI, 2016, pp. 123–135.

Zyskind, G., Nathan, O., and Pentland, A. (2015). "Decentralizing Privacy: Using Blockchain to Protect Personal Data," in 2015 IEEE Security and Privacy Workshops.

6

Hybrid Cloud Adoption in a Developing Economy: An Architectural Overview

Kenneth Kwame Azumah

Centre for Communications, Media and Info. Technologies, Aalborg
University, Copenhagen, Denmark
kka@cmi.aau.dk

Cloud computing in the technological hype cycle has reached the increasing slope of enlightenment. A serious organisation, no matter its size, can no longer afford to ignore the trend that has the potential to increase their business value. It is clear that failure to adopt some form of cloud computing technology today only delays the inevitable in order to stay competitive or be more cost-efficient. The decision to adopt cloud computing however has implications beyond the technical acquisition of an Internet product, raising concerns such as financial impact, organisational readiness, regulatory compliance and security. Depending on the business of the organisation and the geographical context of its operations, the cloud adoption experience could raise further concerns such as Internet connectivity of reasonable speed, cost and reliability, a typical situation sub-Saharan African economies experience. In this paper, the case of a healthcare institution is examined from the perspectives of requirements specification and redesign of an existing network infrastructure to build a hybrid cloud architecture. From the designed constraints, a reasonably successful infrastructure was redesigned together with the testing of selected quality metrics to address cloud adoption concerns of large multi-facility healthcare institutions.

6.1 Introduction

It is common knowledge that the Internet is in operation each hour within each day throughout the year and the Internet is industry works on the premise that there is 99.999% reliability for a piece of information travelling from one computer network to another. This premise underpins the confidence users have in ecommerce and general online business transactions. Because of this implicit confidence, there is a tendency to represent the Internet as a "black box" in relation to the enterprise network or other components. The black box is often shaped as a cloud or nebulous network to indicate its ever-expanding nature. Cloud computing would therefore refer to the computing infrastructure and associated products delivered through the Internet. According to Gartner,[1] adoption of public cloud services are expected to grow to $208.6 billion in 2016 an increase of 17.2% on 2015 where many traditional IT organisations are incorporating their existing datacentres into their overall cloud adoption (Gartner, 2016). In developing economies, cloud implementations are expected to grow, primarily fuelled by high mobile phone penetration rates. For instance in sub-Saharan Africa which has a projected mobile penetrate rate of 51% of total population by 2020 (GSMA,2016)(Hill, 2015) (CISCO, 2016), cloud computing will need to be leveraged to deliver scalable and cost-efficient network architectures.

Distribution of cloud service providers mostly concentrated in developed regions limiting storage locations for consumers from developing economies with storage location sensitive regulations. High bandwidth costs and irregular supply of electrical power makes entry into the market challenging for potential Cloud Service Providers (CSPs) from developing economies. Regulatory compliance enjoins organisations to store data in a known location especially for data national importance or with potential to be affected by international disputes.

These factors induce some of hybrid cloud architecture in organisations seeking to adopt the public cloud but have mission critical applications or sensitive data that must lie within their control. This paper considers the process of hybrid cloud adoption along the perspectives of network architectural redesign to satisfy a given set of requirement specifications taking cognisance of challenges within the context of developing economies. The study derives motivation from the cloud adoption journey of a healthcare organisation and the need to share knowledge with other similar organisations

---

[1]Gartner Inc. (http://www.gartner.com) is an IT related research and advisory firm founded in 1979.

with plans to adopt cloud computing. The selected hospital used as case study implements hybrid cloud computing to accommodate stochastic and bursty workloads seasonally occurring among their clientele. The end-of-month processing of bills, need for patient access to their electronic medical record externally and the need to share same with external collaborators altogether makes the architectural planning and implementation potentially expensive in terms of time and cost.

To measure the success of cloud adoption within the case study, selected metrics and business constraints are analysed in relation to the redesigned network. In Section 6.2, the paper presents a broad overview of cloud computing, its dimensions and state-of-the-art in hybrid clouds. The requirements for designing a computer network are next discussed alongside selected metrics to measure the hybrid cloud. In Section 6.3, the case of the hospital is presented detailing out their network infrastructure and specific components in relation to cloud computing adoption. The needs of the hospital are presented in the form of functional and non-functional requirements of the network redesign. Section 6.4 presents a proposed redesign of the computer network for hybrid cloud adoption, Section 6.5 measures the network performance based on selected metrics, Section 6.6 discusses the results in relation to similar existing solutions and Section 6.7 concludes the various presentations.

6.2 Cloud Computing Background

The term "cloud computing" can be regarded as a metaphor for representing pluggable computing as a utility where the user does not see physical devices but obtains the services from a "black box", the Internet, usually represented as a cloud in diagrams. The concept of cloud computing is not new but rather an old concept enhanced with technology. Parallels can be drawn between today's cloud computing and computing of the 1960's and 70's where time-sharing was used to run multiple jobs at the same time on mainframes or minicomputers. For example, in the community cloud is a semblance of a group of organizations sharing a mainframe in the past few decades.

One of the main technology enhancements that contributed the "sharing" concept was virtualization. Virtualization was introduced to isolate users' jobs running in the time-shared environment and avoid possible errors arising from executing wrong applications or using wrong data. Virtualization separates computing services from the underlying physical environment

(Halper et al., 2012). Cloud computing leverages three characteristics of virtualization: partitioning – where physical resources are divided into portions employing software; isolation – referring to separation of systems where errors or other disruptions are contained within; and encapsulation – where the system is self-contained as a portable single unit (Halper et al., 2012).

Virtualization can be described as having virtual copies of the computing resource, be it hardware or software. The virtualization of hardware is crucial to solving problems of scale, running costs, unused capacity and some form security. In solving problem of scaling, an existing hardware is easily partitioned into copies that can be used independently on demand. For running cost reduction, a virtualized hardware scales at marginal cost due to same hardware consuming same physical resources such as electricity and cooling. For solving unused capacity problems, virtualization creates an avenue for maximizing the use of computing resources for diverse purposes. On issues of security, virtualization allows a server to be compartmentalized hence isolate problems that occur in each virtual machine. A blade server that has most components built onto the motherboard and mounted in a rack in a datacentre was an attempt to isolate server hardware for various users. However, the achievable granularity is limited by the hardware that is fixed and therefore overcoming this by virtualization.

Early virtualization enabled more than one user to use mainframes. However as client computers got more powerful, more processing was moved from the mainframes and data retained for sharing purposes. Later, data was distributed across more servers creating distributed computing where servers were usually distributed over large geographical areas. The development of Hypertext Markup Language (HTML) and Hypertext Transfer Protocol (HTTP) enabled the applications associated with the data to be offered as services, followed by infrastructure as a service fuelled by server virtualization as represented in Figure 6.1. These developments thus gave rise to three main categories of cloud computing Infrastructure as a Service (IaaS), Platform as a Service (PaaS) and Software as a Service (SaaS) referred to as service models.

6.2.1 Service Models

Though there are more categories of cloud computing services rendered, the vast majority fall into the three categories:

- **SaaS:** This model offers users software application through the providers' infrastructure. SaaS has the widest range of services among

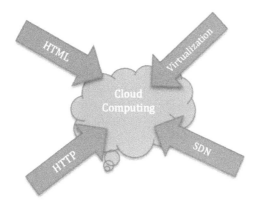

Figure 6.1 Contributing technologies into cloud computing.

the service models because it encompasses all aspects of personal and business software that can be put online. The user consumes the applications as is given and cannot make adjustments to the underlying infrastructure including the operating system or similar platform.

- **PaaS:** In this service model, the user is offered an environment that enables the deployment of custom made application that in turn are consumed by other users. The environment serves as a platform having ready-made plumbing such as databases, blockchain and AI/Machine learning technologies that can be plugged into the deployed applications. The extent of control however stops short of the underlying hardware layer and associated components such as the host operating system and its settings.

- **IaaS:** This service model involves the fundamental computing resource such as virtual machines with RAM, processing power (compute), physical storage and networking. The user is able to provision a wide range of configurations depending on their need. Access is provided through the standard Internet protocols and the user is able to control hardware as much as has been allowed by the provider.

6.2.2 Deployment Models

Another dimension of cloud computing can be viewed from the perspective of how the services are organized, usually referred to as the deployment model. There are four distinct deployment models defined by NIST:[2] public

---

[2]National Institute of Standards and Technology

cloud, private cloud, hybrid cloud and community cloud. The public cloud is accessible to the general public and is open for subscription and access whenever available. It is typically hosted on the premises of the cloud service provider. The private cloud is restricted to an organization where access can only be had by employees or selected collaborators outside the organization. It may be hosted on or off premises of the organization. In the off-premise situation, a dedicated link is put in place to facilitate access. The hybrid cloud is a combination of the private and public cloud where the private cloud can again be on or off premises and the two different cloud deployments linked with a dedicated infrastructure to facilitate the exchange of data between them. The community cloud is a deployment model employed by a group of organizations with similar business operations or interests. The access is limited to these organizations and the deployment may be private, public or hybrid.

Though there are varying definitions of cloud computing the following definition by NIST is one that is widely accepted: *"Cloud computing is a model for enabling ubiquitous, convenient, on-demand network access to a shared pool of configurable computing resources (e.g., networks, servers, storage, applications, and services) that can be rapidly provisioned and released with minimal management effort or service provider interaction."* (Mell & Grance, 2015)

This definition has undergone a number of revisions to address the five essential characteristics that must be present for the cloud-computing paradigm.

- **On-demand self-service**, which is concerned with the value and correct functioning to meet business requirements without human intervention. The Internet is used as the medium of transaction and this also implies there is less concern on the underlying cloud technology.
- **Scalable and elastic**, concerned with the graceful and automatic increase or decrease in level services in response to user demand within reasonable time. Scalability concerns the number of individual instances of the service and elasticity concerns the quantity of resources within the instances.
- **Shared resources**, concerned with a pool of computing resources available for services that are being rendered and offering the higher likelihood of meeting needs on-demand.
- **Measured service**, concerned with tracking and billing of the amount of service utilized by the customer. This also enables several billing models to be designed to suit various user categories.

- **Broad network access**, which employs standard Internet technologies for providing the service to users and also implying the service should be accessible anywhere on the Internet.

There are additional characteristics that are considered essential by Microsoft Corporation but considered as implied in the NIST characteristics (Hu, 2015).

- **Fault tolerance**, concerned with resilience to failures and putting facilities in place to minimize interruptions to the service.
- **Geo-replication**, concerned with having copies of the cloud services strategically located in regions of the world.
- **Security**, for both physical resources and information systems running in the virtual environment.

6.2.3 Considerations for Adopting Cloud Computing in an Enterprise

- **Financial effect:** Operational Expenditure (OPEX) comprises investments spent or exhausted within the year. Capital Expenditure (CAPEX) is usually depreciated over the year and may take a number of years for the capital investment to be exhausted. Both CAPEX and OPEX together constitute the Total Cost of Ownership (TCO) in adopting cloud services. This may also be regarded as the total cost of operating the cloud and consists of direct, indirect as well as operating costs (Gendron & Gendron, 2014). The Return on Investment (RoI) is calculated as a percentage gain on the investment made, offers a rough estimate but worth considering before transitioning to or adopting the public cloud.
- **Readiness of Staff and Users:** One of the crucial factors for successful adoption is the availability of capable Information Technology (IT) staff to administer the new technologies and equally knowledgeable users equipped to employ the newly provided tools in the business operations.
- **Infrastructure availability:** Moving to the public cloud most often entails an increase in bandwidth usage and more bandwidth may be needed if all of the enterprise data is located in the cloud. A key determinant of smooth adoption is the extent to which existing infrastructure can accommodate a new IT paradigm.
- **SLA:** Service Level Agreements are necessary to obtain a level of confidence in cloud services which entire enterprise operations may depend on and over which the IT department may have little to no control. The SLA is therefore a significant determinant in deciding to

adopt the cloud and the provider from whom to obtain cloud services (Gendron & Gendron, 2014) (Waschke, 2012).

- **Core business versus innovation:** The type of cloud to adopt is influenced by the focus of the organization: whether it wants to give priority to its core business operations or channel its resources into innovating within their space. The adoption of the cloud facilitates innovation within the business but also comes with possible recruitment of new staff to manage new technologies. The overall effect becomes a determinant of cloud adoption decision making (Gendron & Gendron, 2014).

Hosted applications in the past have not been built for the cloud paradigm where access and operations have to respond securely to demand without human intervention. Thus corporate datacentres with outward facing applications typically do not have scalability when dealing with RoI or TCO on IT budgets. Hybrid cloud benefits include development on private cloud and deployment on the public cloud. Startups in their business follow the s-curve demands on their resources. They adapt the public, private or hybrid depending on the stage of the s-curve.

6.2.4 Service Level Agreements

SLAs specify the customer requirements in relation to service capabilities (Ramachandran & Mahmood, 2017). It outlines the customer needs and provides a framework for interpreting complex issues that may arise as well as reduce possibilities of conflict. Providers offer to customers standard SLAs that favour themselves and eliminate unrealistic expectations as much as possible. Customers, especially those consuming a lot of services, may opt for a negotiated SLA where metrics are employed to determine and measure the quality of service (Waschke, 2012). Typically, SLA anatomy comprises use of service offerings, security and data policy, customer responsibilities, conditions for temporary suspension, termination conditions, and proprietary rights (Waschke, 2012).

6.2.5 State-of-the-Art in Hybrid Cloud Architecture

The hybrid cloud is a deployment model that combines the public and private clouds reaping the best of both worlds: maintaining operational control in the private cloud and achieving cost-effective computing resource elasticity in the public cloud (Waschke & Waschke, 2015) (Hill et al., 2013) (OMG, 2016).

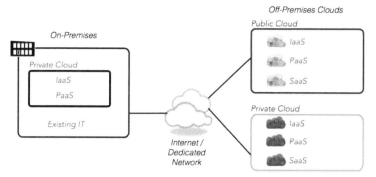

Figure 6.2 Representation of the hybrid cloud architecture.

This is represented in Figure 6.2. Apart from the benefits of the hybrid cloud, the increased complexity of combining two or more individual clouds introduces a number of challenges of overall optimization. A study to analyse the challenges experienced in hybrid cloud adoption revealed performance enhancement, data security and cost optimization as the main thematic areas of concern to aid enterprise leaders in formulating adoption strategies (Azumah et al., 2018). Hybrid cloud architectural components continue to be adjusted in line with research findings in the thematic areas in order to enhance benefits and minimize challenges. The cloud management tool is one such key component undergoing constant development to help address challenges associated with adopting the hybrid cloud and cloud computing in general (Awasthi et al., 2016). A popular cloud management tool, Open-Stack,[3] provides a single pane of glass through which IT administrators can monitor computing resources across the combination of public and private clouds. OpenStack as an open source initiative is constantly increasing in capabilities that encompass various aspects of cloud service provisioning, management and maintenance.

The physical layer of the hybrid cloud consists of components such as datacentre infrastructure and a dedicated link from the public cloud to the organization. Optimizing the hybrid cloud for performance, running costs and security depends on how well the physical layer is organized to suit the workloads of the organization. The advantage of the hybrid cloud easily diminishes with badly made choices in the physical infrastructure that may even introduce security problems. The adoption of a hybrid cloud architecture is therefore done on a case-by-case basis to fit more closely with the business

---

[3] https://www.openstack.org/software/

objective of the organization. These objectives are usually expressed as requirements where the transition to the cloud becomes an IT project having metrics for judging its success or otherwise. The next two sections expand on engineering requirements for hybrid cloud adoption.

6.2.6 Requirements Engineering for the Hybrid Cloud

If the hybrid cloud adoption process does not have the luxury of transitioning wholly to the public cloud due to regulatory requirements, it is essential to find suitable architectures to satisfying business goals instead of relying wholly on the public cloud provider. In traditional system requirements engineering, the architectural design of an information system depends on the requirement specifications. Specifying the requirements of the system to satisfy business objectives entails taking the architectural designs into consideration, especially to arrive at the non-functional requirements. Thus the requirements cannot be specified fully without the architecture and the architecture in turn cannot be designed without the requirements specification. One solution to this chicken and egg problem is via generating a Minimum Viable Architecture (MVA), derived from discussions held between stakeholders of the proposed information system (Koski & Mikkonen, 2017). The MVA is conceived addressing minimum critical functionality for delivering the business goals or overcoming the challenges confronting business operations. The architecture changes to accommodate further requirements as development of the system goes on however the MVA allows some planning to be done towards adopting cloud services for the information system.

6.2.7 Functional Requirements

Functional requirements encompass business processes and there is a high tendency for such requirements to present constraints for the network architecture. To harness the benefits of the hybrid cloud careful planning and transition strategies are employed to avoid extensive interruption in services, ensure there is compliance at every stage of the transition and take cognizance of bandwidth costs arising from large data transfers. The functional view of the network is expressed as layers of function consisting of infrastructure, middleware and software components (McCabe, 2017).

The design goal is to make the network more robust and cost-efficient to operate through functionality such as replication, virtualization and redundancy. As a result, components of the network are grouped into functional

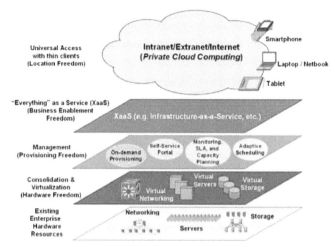

Figure 6.3   Layer-isation of cloud computing modules.

layers that communicate with adjoining layers. Figure 6.3 shows a layer-isation of components that support cloud computing functionality and has the semblance how the computer functions with storage, processing (compute) and network access layers. Each layer's components are selected and organized to address the business goals especially in satisfying constraints placed on the network design. To satisfy the constraint of fast and resilient storage, say, technologies such as RAID and network coding techniques are employed together with fast inter-device fibre connections to facilitate a high level of redundancy. The storage devices themselves should be fast enough employing Solid State Drive (SSD) or flash memory technology to minimize being the bottlenecked point in the network during large data exchanges.

6.2.8 Non-functional Requirements

These are requirements that significantly impact the behaviour of the network in supporting business goals. They encompass quality attributes dependent on the network such latency, throughput and device Central Processing Unit (CPU) utilization but excluding non-negotiable requirements for delivering business value or avoiding potential budget overruns. The quality attributes can be measured employing standard metrics that can be specified in an SLA. NIST defines metric as "a standard of measurement that defines the conditions and the rules for performing the measurement and for understanding the results of a measurement" (NIST, 2014).

Common metric categories in networking measure availability, reliability, response time and throughput.

- **Availability** deals with overall network capacity to mask errors and failures giving semblance of continuous service.
- **Reliability** "refers to the ability to ensure a continuous process of the program without loss" (Bardsiri & Hashemi, 2014) measuring the dependability to recover from failure. Among measures in this category are Mean Time To Recovery (MTTR), Mean Time Between Failures (MTBF) and Response Time Objective (RTO) (Bardsiri & Hashemi, 2014).

  I. MTTR – refers to the amount of time taken recover from failure,
  II. MTBF – refers to the amount of time between failures and
  III. RTO – deals with the duration of system failure

- **Response Time** "…defined as the time it takes for any workload to place a request for work on the virtual environment and for the virtual environment to complete the request". Other names for response time are agility and adaptability (Bardsiri & Hashemi, 2014). This metric directly impacts performance and availability of cloud-based applications.
- **Throughput** "refers to the performance of tasks by a computing service or device over a specific period" (Guiding metrics, 2017). Throughput measures the rate of transport of data in bits per second (bps).

In measuring the above metrics, an independent set of executable and time-bound tasks or code often referred to as the workload, serves as common basis across all categories. Because workloads vary depending on the type and function of the information system (Halper et al., 2012), the appropriate metrics have to be prioritized to achieve business objectives of the network.

6.3 The Case of the Selected Hospital

Healthcare institutions stand to benefit tremendously from implementing information systems that are highly available and yet regulated in compliance with data protection laws. This section describes the business model of a selected hospital in Ghana used as case study. The selected hospital has nine facilities geographical spread across a regional area. The facilities consist of clinics and hospitals with clinics servicing as primary healthcare delivery points and hospital serving as referral centres for secondary and tertiary care.

All facilities are linked via a wide area network to a datacentre accommodated in one of the facilities. The datacentre has four blade (rack-mount) servers and storage systems housed in two cabinets and linked to a backup datacentre accommodated offsite in another branch facility of the hospital. The two datacentres are identical in terms of server and storage capabilities to enable the backup datacentre to operate as an offsite disaster recovery system. In each datacentre, there are four physical servers with multi-core processors and a total RAM of 64GB, 18TB of persistent storage and an additional Network Attached Storage (NAS). The virtual servers generated via a hypervisor share the same physical resources.

6.3.1 The Hospital's Information Systems

One of the virtual servers provides a single relational database management system that is shared among all individual applications running on five other virtual machines covering Electronic Medical Records and Billing (EMR), Human Resource and Accounting Information Systems (HRAIS), the Radiology Information System (RIS), Pharmacy and Laboratory Information Systems (PLIS), and a VoIP gateway. A major IT strategy of the Hospital is to focus on integrating the various systems, reduce the tedium of data entry and thereby improve overall efficiency of the dependent healthcare services. For example, an integration EMR system and HRAIS facilitates automated aggregation and entry of patient bills as receivables in the chart of accounts. Similarly, the PLIS, EMR and RIS are all integrated to offer a complete history and billing overview on patients. The integration and communication between the various systems are both loosely coupled via APIs and tightly coupled via direct code in application. This places a constraint on any cloud transition process to ensure tightly coupled applications are placed together within a datacentre to minimize possible latency problems or avoid high bandwidth costs between applications that are chatty in nature.

6.3.2 Hospital Network Requirements and Business Constraints

Increasingly, the dependency of the Hospital's information systems on the network gets complex whenever new clinical operations and services are introduced. For example, the introduction of a higher resolution CT Scan or other medical imaging device places pressure on the bandwidth within the wide area network and any effort to transition the RIS to the cloud has to take cognizance of the available Internet bandwidth or risk degrading the users' experience. The network requirements for adopting a hybrid cloud architecture are considered from the functional and non-functional perspectives whilst the business constraints focus on general non-negotiable requirements of compliance, cost-effectiveness, access and security.

6.3.3 Functional Requirements Specification of the Hospital Network

- Physical network components should be modular, commercially available and hot swappable as much as possible. Modular parts should be organized in server, storage, datacenter networking fabric, WAN and Wi-Fi[4] LANs.
- The network should be able to carry voice data with ease.
- The virtual servers for medical and healthcare related information have to reside on the same physical server. Similarly, information for hospital administration and communication should be put together on a physical server.
- All computing resources in both datacenters should be linked with high-speed infrastructure and accessible via a cloud management tool (cloud operating system). The management tool should be web-based and be able to handle inventory, address, routing and name management, network discovery, operating system image upgrades and distribution, IT staff authentication and finally capable of distributed intelligence.
- The database server having personally identifiable information should always reside in within the internal datacenter and should only accept encrypted connections.
- Provision should be made for caching and load balancing to handle peak and bursty workloads.
- The design should provide access for external collaborators, contracted surgical consultants and patients outside the Hospital premises.

6.3.4 Non-functional Requirements of the Hospital Network

- The design is to have an overall availability of 99.50%.
- The combination of metrics comprising MTTR and MTBF on the network should be able to support the availability threshold.

---

[4]Trademark phrase for IEEE 802.11x

- Throughput should not be less than 5Mbps and 10Mbps for peak and off-peak times.
- Response time metric should support the throughput objectives.

6.3.5 Other Business Constraints

Workloads determine the strategy of cloud adoption; multi-server applications can be transitioned into the cloud. Intensive compute applications that run periodically and stochastic compute applications that are more event-driven may be run in the internal datacentre or public cloud. Increasingly off-premise private clouds are offering scalability features that are regulation compliant such as HIPAA. This offers more advantage over on-premises private cloud in terms of cost-savings arising from economies of scale at the providers' datacentre among which is cost-shared cooling, electricity and physical security.

6.4 Infrastructure Redesign and Proposed Solutions

The aim of the IT project is to prepare the network for transitioning information systems partly into the public cloud and maintaining critical applications and sensitive data in the private datacentre. The first most significant change to be done in the existing network comprises the building and configuration of dedicated VPN infrastructure into the preferred public cloud. The link must be backed by multiple Internet service providers in order to introduce the necessary redundancy. A Virtual personal Network (VPN) manager at both the public cloud and internal datacentre route requests from the public cloud to the private datacentre and responses from the private datacentre to the public cloud.

A second redesign goal is the replication of the master database system to facilitate high availability to applications in the load balancer. In Figure 6.4, the Domain Name System (DNS) server routes incoming traffic to the load balancing virtual machines in the public cloud. The load balancer in turn distribute the traffic among the various application servers, more of which are spawned automatically by the hypervisor up to a limit agreed in the provisioning agreement or SLA. The application servers hit the database via a VPN tunnel, which runs through the Internet. A further option in the case of the off-premises private datacentre is the direct access to database within the premises of the provider instead of a VPN tunnel through the Internet which may be prone to latencies.

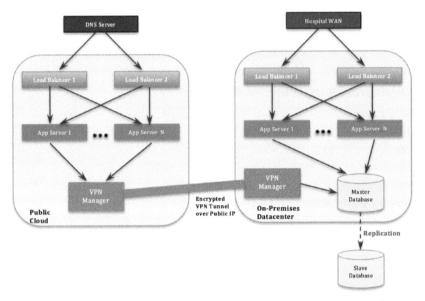

Figure 6.4 Hybrid Cloud to enhance availability and ensure regulatory compliance.
Source: (RightScale, 2018)

Requests from the hospital WAN are also managed by load balancers inside the internal datacentre which in turn distribute incoming workloads among virtual machines running the various hospital applications. In the internal datacentre, care must be taken in spawning new virtual machines due to a constraint of limited computing resources, mainly RAM and processing time (compute). In processing the workloads the application servers directly access the database located within the same physical infrastructure but separate logical space to isolate from possible application processing bottlenecks and errors.

In a backup datacentre located offsite, a slave database replicates the master on a regular basis through a high-speed link on the WAN. This replication principle can be reproduced in each hospital facility to bring the database close to LANs within the WAN, reduce access times whilst maintaining a high data-sharing capability.

The cloud operating system OpenStack manages the entire hardware infrastructure. It features functionality for processing virtual machines, storage, memory and networking across both public and private clouds all in a singe pane of glass.

6.5 Measuring Performance Results

This section offers basic analyses to determine satisfaction of metrics specified in the requirements and looks into the general organization of the network. The analysis employs the metrics as guidelines to judge the appropriateness of the infrastructure. Monitoring the network done over a period of two months produces measurements such as MTTF, MTBF and MTTR which give an insight into quality attributes of availability, reliability, throughput and response time.

The measurements employed tools such as the Wireshark network protocol analyser and Telerik Fiddler, a web-debugging tool respectively for the selected metrics of throughput and response time. Monitoring over the two-month period saw an average of four minutes downtime in some part of the network every four days resulting in an overall availability of 99.9306%.

$$Availability = \frac{MTBF}{MTBF + MTTR} = \frac{4\ days}{4\ days + 4\ mins}$$

$$= \frac{5760\ mins}{5764\ mins} = 99.9306\%$$

The throughput in the hospital network requirements specification ensures there is adequate bandwidth to transfer radiological and other medical images at reasonable speed. Figure 6.5. gives the average response times from an application in the radiology department. It shows 1,092,692 bytes transferred in response to a request and calculated from the moment the server began a response to the time the client completed receiving the response: 4.432 − 3.818 = 614 milliseconds. This gives a throughput of 1.69 Mbytes per second:

$$Throughput = \frac{data\ size}{transfer\ time} = \frac{1,092,692 Bytes}{0.614\ seconds}$$

$$= 1.69\ MB/s$$

In practice however, transferring a medical image from the RIS across the WAN achieved an average throughput of 1.35 Mbytes per second.

```
Request Count:      1
Bytes Sent:         919          (headers:919; body:0)
Bytes Received:   1,092,981          (headers:289; body:1,092,692)

ACTUAL PERFORMANCE
--------------
ClientConnected:        04:33:01.208
ClientBeginRequest:     04:34:02.408
GotRequestHeaders:      04:34:02.408
ClientDoneRequest:      04:34:02.408
Determine Gateway:      0ms
DNS Lookup:             0ms
TCP/IP Connect:         0ms
HTTPS Handshake:        0ms
ServerConnected:        04:33:33.027
FiddlerBeginRequest:    04:34:02.408
ServerGotRequest:       04:34:02.412
ServerBeginResponse:    04:34:03.818
GotResponseHeaders:     04:34:03.818
ServerDoneResponse:     04:34:04.432
ClientBeginResponse:    04:34:04.432
ClientDoneResponse:     04:34:04.432

         Overall Elapsed:      0:00:02.024

RESPONSE BYTES (by Content-Type)
--------------
text/html: 1,092,692
~headers~: 289
```

Figure 6.5   Response times as measured with fiddler.

6.6 Results Discussion

This section presents a discussion on the network redesign results in relation to existing literature. It discusses the standard hardware and software components employed to (i) connect a private cloud to the public cloud and (ii) enhance interoperability whilst satisfying the given business constraints. This section also discusses the metrics and their monitoring roles in maintaining a desired QoS within the cloud architecture.

Teckelmann et al. (2011) present a taxonomy of interoperability for IaaS among which are (i) access mechanism, (ii) virtual appliance management, (iii) storage, (iv) network and (v) SLA. Access mechanism uses HTTP REST API for both discovery and consumption of available services. The network redesign employs the same hypervisor both in the private and public cloud. This allows the migration of Virtual Machines (VM) between the two clouds, reduces the setup time in making them operational and supports high availability of the software stack hosted in on the VMs. Maintenance of the VMs take place via OpenStack which provides a single window through which both public and private cloud VMs may be administered. OpenStack also provides a management interface for storage of VM images, snapshots, backups and replication. The storage is decoupled from the VMs to enhance interoperability of the two clouds. In the network redesign, the replication of the master database into a backup datacentre supports the high availability requirements and facilitates maximum control over data storage location and access.

For network interoperability, Teckelmann et al. (2011) mention two areas: addressing and application-level communication. The case study hybrid cloud implementation employs standard IPv4 and IPv6 addressing schemes managed through the OpenStack cloud operating system. Inter-virtual machine communication is primarily via Internet Protocol and various addressing techniques help maintain a connection to the VMs during migration between subnets of a cloud. IPv6 address solves the potential IPv4 address shortage however adoption of IPv6 addressing as mainstream has been slow owing to subnetting which allows extensions to existing the IPv4 pool of addresses. HTTP facilitates application-level communication making use of mechanisms including RESTful APIs and SOAP. Communication in the redesigned network employs well-established protocols and mechanisms in the Internet Protocol and HTTP contexts.

The security aspects of the cloud internetworking focus on authentication, authorisation, accounting and encryption as mechanisms to enhance trust and privacy in cloud data exchange and communications. User management is a central feature for the infrastructure employing single sign-on (SSO) techniques to grant access to a range of services available in both private and public clouds. The encryption is an essential feature in securing communication and sensitive data in transit.

Event-driven logging facilitates accounting and traceability, and serves as input records for overall infrastructure health monitoring tools. For interoperable authentication mechanisms, the Public Key Infrastructure (PKI) provides the extensible foundations that can incorporate certificates for authenticating devices and services.

Toosi et al. (2014) present three approaches of interoperability between clouds: (i) use of standard interfaces, (ii) employing middleware and (iii) employing a combination of both. The motivation for OpenStack is quickly becoming a de facto standard with massive efforts to improve interoperability and harness the benefits enumerated by Toosi et al.: increased resource availability, minimized danger of vendor lock-in, more efficient geographic distribution and access, increased potential for regulatory compliance and enhanced cost efficiency.

In answer to functional requirements, one of the major goals was to ensure modularity and interoperability, facilitate easy maintenance and avoid vendor lock-in. From the redesign, the array of application servers provide modular arrangements where VMs can be moved between the public and private clouds to address high and unanticipated demand. In shifting VMs between the clouds, the design takes cognisance of high-speed bandwidth availability through an encrypted VPN tunnel to minimise latencies in VM migration and restarting.

Load balancing is a noteworthy network component that caches repeated data requests to further reduce the latencies and burden on application servers. In the case of the hospital, this is especially useful in look-up data access for external collaborators and patients interaction with their medical records outside the hospital network.

In addressing the non-functional requirements, overall availability achieved a satisfactory level registering 99.93% against a laid out threshold of 99.50%. Most often, the desired availability for a hospital is 99.99% culminating in one minute downtime per week or 52 to 53 minutes spread throughout the year. A lower threshold than the four 9s, for the selected case of the hospital, usually arises from the erratic power supply characteristic of the regional context of Ghana. Networks are normally supported by Uninterruptible Power Supply (UPS) systems however the repeated power surges or fluctuation many times causes failures.

The selected metrics belong in the general category as classified by Bardsiri and Hashemi (Bardsiri & Hashemi, 2014). They offer a quantitative approach to monitoring and measuring service quality. In the SLA, a combination of metrics measures the network performance and provides objective ways of accessing the relationship between business Key Performance Indicators (KPIs) and the quality of network performance. SLAs in this sense become monitoring tools for maintaining desired thresholds in support of key performance indicators that may vary from year to year or from department to department within the organisation. The ideal or desired performance of the hybrid cloud infrastructure is state in the SLA whilst the KPIs measure the true or current performance with reference to the selected metrics. In the case of the Hospital, availability and throughput are key requirements determined through the combination of selected metrics MTTR, MTBF and MTTF. Another metric dealing with mission-critical operations is the mean time between critical failures, MTBCF, which helps measure availability of mission critical hours. McCabe, 2017, presents typical requirements for network analysis and maintenance: RMA standing for Reliability, Maintainability and Availability. The metrics culminating in their measurement are: for reliability, MTBCF which is usually expressed in hours; for maintainability, MTTR which statistically measures the time to fully restore an occurring system failure; and for availability, the relationship between MTBCF and MTTR measures the frequency of breakdowns during mission-critical hours.

6.7 Conclusion

Transitioning an existing information system partly into the public cloud entails both technical and user preparation. The challenging aspects lie in the areas where there is a least amount of control, in the public cloud and in user behaviour. To plan the transition effectively, some requirements engineering is needed however, in cloud adoption this is a complex endeavour mainly due to the uncertainties in reliability, availability and changing regulatory requirements. In the hybrid cloud, the level of complexity increases because of the combination of two or more different cloud deployment models, typically the public and a private cloud. For addressing issues of maintaining control whilst exploiting scalability, a hybrid cloud consisting of a private cloud tethered to public cloud provisioning is highly desired especially by organisations that already have an on-premise datacentre running some sort of virtualisation systems. The on-premises private cloud provides the needed control of sensitive data and applications to comply with regulatory requirements and data protection laws especially where personally identifiable information is concerned. The preceding sections looks at the redesign of an existing infrastructure to facilitate hybrid cloud process and minimize disruptions to the hospital operations, used as case study. The study focuses on technical preparations needed to achieve a set of specified requirements. In redesigning an existing network, the set of functional and non-functional requirements provides specifications that support the core operations of the organisation and satisfy peculiar constraints such as cost or bandwidth limitations. Metric-based specifications for the negotiable non-functional requirements facilitate measurement and monitoring on a regular basis. The case employed in this study is one of a hospital in Ghana that has nine facilities spread over a geographical area and linked in a WAN. The common on-premises data storage safeguards data-sharing between the facilities and the network redesign facilitates more control in handling heavy workloads via public cloud provisioning and an encrypted VPN. The implemented network redesign meets the requirement specifications in terms of SLA quality attributes of reliability, maintainability and availability. Metrics aid in measuring set out quality attributes and provide quantitative measures for monitoring against thresholds in SLAs. Existing related works deal with network analysis and design, cloud adoption requirements engineering and interoperability issues with IaaS. From the related literature communication between clouds and their components via APIs facilitates management of the services being provided. A management tool that is increasing gaining prominence is OpenStack, providing a single pane of glass for managing networking, storage, compute, billing, user and virtual machine management. The implemented case of the hospital provides a good example of how existing network design and architecture literature support hybrid cloud computing and provide knowledge for transitioning from an on-premises infrastructure towards adopting public cloud provisioning whilst satisfying a given set of business constraints. The measurement section showed the example of measuring the selected metrics of availability and throughput. The redesign section discussed the new network topology after adopting hybrid cloud architecture. It showed satisfaction of business constraints such as access to internal datacentre via the public cloud and a strong workload distribution infrastructure. Measurements indicated a satisfactory performance in availability and throughput over a two-month period however these metrics need constant monitoring for a longer period of time to be certain of network redesign satisfying the given business constraints.

References

Awasthi, S., Pathak, A., and Kapoor, L. (2016). Openstack-paradigm shift to open source cloud computing & its integration. In *2nd International Conference on Contemporary Computing and Informatics (IC3I)*, pp. 112–119.

Azumah, K. K., Sorensen, L. T., and Tadayoni, R. (2018). Hybrid Cloud Service Selection Strategies: A Qualitative Meta-Analysis, in *2018 IEEE 7th International Conference on Adaptive Science and Technology (ICAST)*, pp. 1–8.

Bardsiri, A. K. and Hashemi, S. M. (2014). QoS Metrics for Cloud Computing Services Evaluation, *Int. J. Intell. Syst. Appl.*, 6(12), pp. 27–33.

CISCO. (2016). Cisco Global Cloud Index: Forecast and Methodology, 2015–2020.

Gartner, (2016). Gartner Says Worldwide Public Cloud Services Market to Grow 17 Percent in 2016. Available: http://www.gartner.com/newsroom/id/3443517. [Accessed: 31-Dec-2016]

Gendron and Gendron, M. S. A History of How We Got to Cloud Computing, in *Business Intelligence and the Cloud*, Hoboken, NJ, USA: John Wiley and Sons, Inc., 2014, pp. 3–21.

GSMA, (2016). GSMA Mobile Economy 2016. Available: http://www.gsma. com/mobileeconomy/. [Accessed: 20-Dec-2016].

Guiding metrics, (2017). Cloud Services Industry's 10 Most Critical Metrics – Guiding Metrics. Available: http://guidingmetrics.com/conte nt/cloud-services-industrys-10-most-critical-metrics/. [Accessed: 20-Feb-2017].

Halper, F., Kaufman, M., and Hurwitz, J. (2012). *Hybrid Cloud For Dummies*. Wiley.

Hill, K. (2015). Cloud computing emerging in Africa – RCR Wireless News. Available: http://www.rcrwireless.com/20151023/featured/cloud-computing-in-africa-tag6. [Accessed: 02-Jan-2017].

Hill, R., Hirsch, L., Lake, P., and Moshiri, S. (2013). *Guide to Cloud Computing*. London: Springer London, 2013.

Hu, T.-H. (2015). *A Prehistory of the Cloud*.

Koski A. and Mikkonen, T. (2017). What We Say We Want and What We Really Need: Experiences on the Barriers to Communicate Information System Needs, in *Requirements Engineering for Service and Cloud Computing*, Cham: Springer International Publishing, pp. 3–21.

McCabe, J. D. (2007). *Network analysis, architecture, and design*. Burlington: Elsevier/Morgan Kaufmann Publishers.

Mell, P. M. and Grance, T. (2011). The NIST definition of cloud computing, *NIST Spec. Publ.*, 145, pp. 7.

NIST Cloud Service Metrics Sub Group, (2015). Cloud Computing Service Metrics Description.

OMG: C. (2016). Cloud Standards Customer Council, Practical Guide to Hybrid Cloud Computing.

Ramachandran M. and Mahmood, Z., Eds., (2017). *Requirements Engineering for Service and Cloud Computing*. Cham: Springer International Publishing.

RightScale, (2018). Cloud Computing System Architecture Diagrams. Available: http://docs.rightscale.com/cm/designers_guide/cm-cloud-computing-system-architecture-diagrams.html.

Teckelmann, R., Sulistio, A., and Reich, C. (2011). A Taxonomy of Interoperability for IaaS, in *Cloud Computing*, CRC Press, pp. 45–71.

Toosi, A. N., Calheiros, R. N., and Buyya, R. (2014). Interconnected Cloud Computing Environments, *ACM Comput. Surv.*, 47(1), pp. 1–47.

Waschke, M. (2012). *Cloud Standards*. Berkeley, CA: Apress.

Waschke, M. (2015). *How clouds hold IT together.: integrating architecture with cloud deployment*. Apress.

7

Developing Use Cases for Big Data Analytics: Data Integration with Social Media Metrics

Ezer Osei Yeboah-Boateng[1] and Stephane Nwolley, Jnr[2]

[1]Ghana Technology University College (GTUC),
Ghana and EZiTech, Ghana
[2]Npontu Technologies, Ghana
eyeboah-boateng@gtuc.edu.gh, snwolley@gmail.com

The magnitude of data generated and shared by various stakeholders has been exacerbated due in part to social media networks and platforms. These datasets emanate from disparate sources, with varied formats, such as unstructured, semi-structured and structured, and in a variety of types, including multimedia, textual contents, etc. Today, various entities (i.e. humans, computer systems, devices, firms, etc.) generate these Big Data, which are in huge volumes, and requiring non-traditional database management technologies for storage and processing. Some concerns exist with the handling, integration and processing of the datasets, as they come from diverse sources. The essence of data integration is such that an event that occurs in one social media metric would automatically trigger an event in one or more separate metrics. The approach adapted was site scraping to collect datasets, and system design principles applied to conceptualize a solution. Key contribution is the social media metrics computational and virtual data integration model aimed at assisting stakeholders as a guide in gathering, processing, analyzing and utilizing insights for competitive advantage. Typical outcomes are innovative trends, patterns and even, outliers. The outliers may be hidden novelty that ought to be explored further. SMEs in developing economies are likely to find this study useful. Further research may experiment with actual social media metrics.

7.1 Introduction

There has been more data captured and processed over the last two years than there has been in the whole world's existence. It may not come as surprise considering the various sources of data generated and collected during the latter part of the 20th century. The magnitude of data generated and shared by businesses, public administrations, numerous industrial and not-for-profit sectors, and scientific research, has increased immeasurably (Agarwal & Dhar, 2014). These data include textual content (i.e. structured, semi-structured as well as unstructured), to multimedia content (e.g. videos, images, audio) on a multiplicity of platforms.

In our current stage of globalization, there exist various channels like social media, cell phones, Global Positioning System (GPS) signals, online payment platforms, online mailing platforms, chat platforms, online research centers and much more which are generating, collecting, storing and using information from more than 1 billion people at a time. Every day the world produces around 2.5 quintillion bytes of data (i.e. 1 Exabyte equals 1 quintillion bytes or 1 Exabyte equals 1 billion Gigabytes), with 90% of these data generated worldwide perceived to be unstructured (Dobre & Xhafa, 2014). By 2020, over 40 Zettabytes (or 40 trillion Gigabytes) of data will have been generated, imitated, and consumed (Gantz & Reinsel, 2013).

The complexities in the data being collected and the sheer volume of interaction and processing between people and devices is heretofore being referred to as the Big Data. "Big Data refers to a combination of an approach to informing decision making with analytical insight derived from data, and a set of enabling technologies that enable that insight to be economically derived from, often, very large and diverse sources of data" (Akred, 2014). Big Data as a data represents a cultural shift in which more and algorithms with transparent logic, operating on documented immutable evidence, make more decisions.

IBM most appropriately in the earlier stages of the definition of Big Data introduced what was referred to as the 3Vs of Big Data and today these V's have been expanded over 12Vs. In 2001, Gartner's Doug Laney first presented what became known as the "three Vs of big data" to describe some of the characteristics that make big data different from other data processing. "Big data as a data that, by virtue of its velocity, volume, or variety (the three Vs), cannot be easily stored or analyzed with traditional methods" (Poulson, 2014). As the world enjoyed its newfound status of globalization, making the globe smaller than intended with the various levels of interactions created, the convolution of information generated could crash any machine resulting in a more sophisticated manner of data collection and processing. Thus, now terms like Gigabytes, Terabytes, Petabytes, Exabytes and Zettabytes have graced our computing language when one speaks of data storage. The development of the analysis of Big Data revealed that the mass of data received in its raw form was of 3 main formats and as mentioned briefly above, these formats were: *Structured, Semi-Structured and Unstructured;*

- **Structured:** the reference to a structured form of data represents a variety of information, collected and organized using relational entities. "Structured data refers to information with a high degree of organization, such that inclusion in a relational database is seamless and readily searchable by simple, straightforward search engine algorithms or other search operations" (Planet, 2012). Structured data is information, usually text files, displayed in titled columns and rows which can easily be ordered and processed by data mining tools.

    In a Search Engine Optimization (SEO) context, "structured data usually refers to implementing some type of markup on a webpage, in order to provide additional detail around the page's content" (Randolph, 2017).

This makes for easy identification and usage. There are a variety of relational entities as there is information. These entities could be: age, gender, size, number format, color etc. Such unique identifiers presented by the various relational entities present an easily recognizable data that does not need much refining. The data management system storing the information is set up to understand and interpret any information inputted using the defined relational entity described, with strict requirements on appropriate data table storage. Structured Data does not present as many challenges in data processing as other forms of data do, but often data collected is more unstructured than structured. The programming language in the Relational Database Management System (RDBMS) is Structured Query Language (SQL). The management responsibilities for DBMS encompass the information within databases; the processes applied to databases such as access and modification; as well as the logical structure of the database (Raza, 2018).

- **Semi-Structured:** It is usually said that the line between unstructured and semi-structured data is a thin one. This is because although data in this place does not follow a strict data format required for storage in a relational database, it's neither varied enough to be called unstructured, nor organized enough for structured. Instead, the semi-structured data is stored in a distinctive data space other than a relational database, using tags that act as unique identifiers which enable identification, interpretation and usage. Semi-structured data as a type of data that contains semantic tags, but does not conform to the structure associated with typical relational databases (Raza, 2018). Semi-structured data is rather popular now with the advent of the Internet causing the rise in the need for languages like XML and JSON to facilitate the matching of common data presented in different data styles; e.g. bibliography data.

- **Unstructured:** One thing that can be said to be common to both structured and unstructured data is its ability to be recognized, and its provision in a predictable file format. Unstructured data obeys no rules whatsoever and presents no systemic format for data querying. Unstructured data is even more popular these days in the wake of social and multimedia platforms. Unstructured data continues to grow in influence in the enterprise as organizations try to leverage new and emerging data sources. Unstructured data is any data that aren't stored in a fixed record length format, which is known as transactional data (Shacklett, 2017). These new data sources are made up largely of streaming data coming from social media platforms, mobile applications, location services, and the Internet of things (IoT) technologies (Hahn, 2017).

As data gets bigger, more ubiquitous, and more social, some of these data may be vital or critical data needed for business decision-making. Somehow, interactions with social media platforms must, ought and should create value for the business (Yeboah-Boateng, 2019). The question is, how does the business harness these data? How does the business integrate the disparate data sources to create trusted data stream or actionable information. The trick is to identify specifically what social data is relevant and to analyze just exactly what that data means. There are a few of metrics, tools, and services to help businesses make sense of its social data, as with any other data being managed. Some general rules are; to seek out more unique metrics, and to avoid overemphasizing simple counts and totals.

Social media marketing includes so many topics and tactics such that measuring performance often can prove elusive. Still, just as is the case with any other marketing initiative, one needs to focus on the activities that deliver the greatest benefit (Lovett, 2011). Here, we highlight five of the social media metrics to buttress our point of view in the study. These are Volume, Reach,

Engagement, Influence, and Share of Voice. By employing the agile methodology and scraping, we gleaned through extensive literature of key articles, online databases, and authorities, towards improving the data integration with Big Data use cases and sharing approaches that may be easily adaptable by SMEs in developing economies (Yeboah-Boateng, 2013a) (Yeboah-Boateng, 2017). Typically, the agile development model requires extreme agility with benchmarking the metric for new media and its associated computations. This study is an exploratory survey of Big Data and value creation aspects of knowledge management practices employed in new media. Sources of data have been employing use cases of domestic, transnational and global literature on Big Data value creation focusing on new media.

From the perspective of system development and analysis, the study makes contribution with a conceptual use cases and corresponding data integration model aimed at harnessing Big Data based social media metrics.

7.1.1 Problem Formulation

Big Data is characteristic of huge datasets generated at fast rate, in unstructured, semi-structured and structured formats, emanating from disparate sources (Yeboah-Boateng, 2019). Social media is a massive contributor to these increasing volumes of data which could be harnessed to transform the data from measures and metrics into usable information, and to create the necessary critical data for business decisions (Lovett, 2011). The extent of data volumes generated through social media, sensors and actuators, log files, etc., has fueled the emergence of Big Data. Indeed, the sheer volume, velocity and variety of data generated far outstrips traditional data warehousing capabilities (Datameer, 2016). It must be noted that the value and insights are not necessarily obvious, except through Big Data analytics designed to harness ("unlock") them with "specific use cases and applications" (Datameer, 2016, p. ii).

Technically, due to disparate sources containing these datasets, appropriate data integration approaches ought to be designed and deployed to harness the insights and value for business use. Datameer (2016) in its report identifies five (5) top use cases aimed at harnessing Big Data to deliver the highest value.

It is imperative that organizations explore innovative use cases to harness the unprecedented value to leverage sustainable competitive advantage (Yeboah-Boateng, 2019) (Datameer, 2016).

The social media craze is fueled by human nature; i.e. as social creatures who communicate their "needs, wants and desires" (Lovett, 2011, p. 30) to one another. In this digital era, these interactions from mere social gestures to audio/sound bites are, through the digital media, recordable, stored and analyzed. These actions performed using social media platforms; i.e. from clicking, reviewing, commenting, reading, - all contribute to the Big Data.

The essence of deducing valuable actionable information from the datasets is not in doubt. Businesses create value from these Big Data. However, trust is a key factor in engagement on social media; e.g. use real pictures as profile image, not pets or company logos, etc. Various concerns are raised in respect of the generated datasets; are they coming from trusted and reliable sources? Are the insights derived representative of the factors, or opinions, or observations? What processes are used to integrate the data from the disparate sources? How do we ensure that the realistic and relevant metrics are employed by the organization in decision making? What are the use cases necessary to harness trusted data integration?

In this study, we concentrate on how to make sense of the data generated for value creation (Yeboah-Boateng, 2019). The usual functional requirements listing is often static and falls short of the needed business processes to be supported.

7.1.2 Key Research Questions and Objectives

Key research questions and objectives for this study are presented hereunder:

- Which data points are critical to the Big Data analytics?
- How do we harness insights from data integration from trusted sources?
- What use cases are derived from the integrated social media metrics?

The objectives of the study are as follows:

- To design a social media metrics algorithm to harness value creation for organizations;
- To assess some social media metrics used in benchmarking Big Data analytics;
- To develop some use cases needed to enhance social media metrics, using Big data analytics.

7.1.3 Highlights of Findings

Key findings are the catalogue of computational equations used in benchmarking the social media metrics utilized by corporate executives for competitive advantage. Similarly, an intuitive strategic Big Data based Social Media Metric Computational and Data Integration model aimed at assisting stakeholders as a guide in gathering, processing, analyzing and utilizing actionable information that could be harnessed to formulate appropriate competitive advantage strategy is proposed.

7.1.4 Significance of the Chapter

The study's appeal with perspectives from Big Data analytics, social media metrics and data integration – from datasets generated, captured and processed from disparate sources, integrated, analyzed and utilized, to create value. It presents some business use cases and scenarios, as illustrations of the key concepts. The notion of virtual data integration is applied to ensure that the datasets are collated and combined appropriately for processing to harness the invaluable insights for competitive advantage.

7.1.5 Outline of the Chapter

This introductory section dealt with overview of Big Data and new media aspects of knowledge management generation and utilization. Followed by related works on knowledge-based modeling, Big Data, new media metrics and datasets, approaches employed in processing Big Data generated. Agile techniques employed to source data are discussed. By intuitive deductions and critical thinking the model is designed and its implications for strategic value creation based on social media metrics computations and virtual data integration follows, and then conclusion and recommendations.

7.2 Literature Review

This section deals with the key thematic areas of the study. It commences with the use cases and business scenario modeling, dealing with social media domain-specific facts and rules as the fundamental metrics upon which the conceptual data integrated social media metric model is built. Then, the concept of data integration is discussed in respect of challenges and prospects and its virtual architecture. Various approaches necessary for dealing with Big Data are also discussed. The concept of social media metrics follows. Social media use cases, with case examples are discussed. The foundational social media metrics that organizations use to measure the value contributions of social media are also discussed extensively.

Finally, value creation strategies, with emphasis on social media generated Big Data, are presented.

7.2.1 Business Use Cases

Use cases are intended objectives or intended system behavior, used in system design analysis to identify, clarify and classify system functional requirements. They indicate the possible system interactions between end-users and the system.

The end-users, often referred to as actors, could be human users, computer systems, organizations, software systems, etc.

Typical use case, according to (Swafford, 2006) consists of the following:

- Use case name;
- Actors
- Summary;
- Pre-conditions;
- Flow of Events;
- Error Conditions;
- Post Conditions.

The use case is used to depict the desired functionality, the constraints imposed and the business objectives to be fulfilled. In this study, we employ use cases to uncover implied functionalities needed by business leaders to create value from Big Data based social media metrics. Additionally, use cases offer a great deal of insights into the appropriate social media metrics needed by the organization to optimize its value; i.e. to create the desired competitive advantage.

Use cases are specific business objectives that an intended system needs to accomplish. Use cases describe the processes of external actors or entities that exist outside the system, as well as the specific interactions they perform to achieve the desired business objectives. Business use cases, also known as Abstract-Level Use cases, are typically technology agnostic, and depict the sequence of actions required to perform or achieve some specific business objectives.

In this study, we model a holistic business use cases analytics together with some business scenarios (or what is known as user story). The business scenarios are typically narratives that describe how users interact with the system functionally. They provide descriptions of the tasks to be carried out to accomplish the desired business objectives.

The proposed data integrated business use case model depict discrete use cases or goals or objectives, based on UML modeling principles.

7.2.2 Data Integration Principles

Data integration is the combination of technical and business processes used to combine data from disparate sources into meaningful and valuable information. A complete data integration solution delivers trusted data from various sources.

Data integration is positioned to handle external sources of information, with associated distributed nature of the Web and issues emanating from information sharing; especially, uncertainty of data generated. Data integration is:

- To facilitate manipulation of heterogeneous data classification (matching);
- To facilitate data extraction from various Web sources (e.g. social media sources) and for storage of integrated data;

The principle of data integration is such that an event that occurs in one social media metric module would automatically trigger an event in one or more separate metric modules. The essence of the data integration, i.e. combining data from sources across multiple social media sources, can afford businesses (and business leaders) opportunities to obtain competitive advantage and for improvement. Data integration facilitates harnessing knowledge (data) to get insights into products and services offered by the company.

Data integration is a key challenge in social media metrics, especially where interactions occur in real-time, with diverse unstructured data sets or feeds that ought to be harnessed for insights about a company's products and services. It also finds applications in the decentralization agenda of African governments, as they seek to coordinate activities of different Metropolitan, Municipal and District Assemblies (MMDAs) in its digital transformation pursuits; e.g. African Union (AU) Agenda 2063. Typical characteristics for data integration are (Doan et al., 2012):

- Query – querying and updating disparate data sources;
- Number of sources – addressing challenges in managing diverse data sources for trusted data;
- Heterogeneity – variety of data formats of social media metrics (e.g. unstructured, structured, and semi-structured data sources);

- Autonomy – ability of social media metrics to change with time, without due reference to prior formats and access patterns.

In terms of systems architecture, this study is hinged on a virtual data integration, such that the social media datasets (metrics) remain in the original sources and are accessed as and when needed at query time (Doan et al., 2012, p. 9).

Data integration is the process of consolidating data from a variety of sources in order to produce a unified view of the data. It enables the following process flows to accomplish the tasks (SAS Institute, Inc., 2007):

- To Extract, Transform and Load (ETL) data for use in data warehouses and data marts;
- To cleanse, integrate, synchronize, replicate and promote data for applications and business services.

7.2.3 Big Data Analytics

Big Data is often perceived as datasets that are so large and complex, such that traditional tools for storage, processing and analysis are unable to process them "in an acceptable time frame or within a reasonable cost" (Frampton, 2015) (Erl et al., 2016).

There are some concerns with social media based Big Data, in respect of sourcing, moving and analyzing data. Virtual data integration techniques may be employed to address some of the issues observed. Frampton (Frampton, 2015, p. 3) posits that the algorithm must necessarily follow the flow processes enlisted hereunder:

- A method of collecting and categorizing data;
- A method of moving data into the system safely and without data loss;
- A storage system that is distributed across many servers;
- A rich tool set and community support;
- ETL-like tools (preferably with a graphic interface) that can be used to build tasks that process the data and monitor their progress;
- Scheduling tools to determine when tasks will run and show task status;
- The ability to monitor data trends in real time;
- Local processing where the data is stored to reduce network bandwidth usage.

The definitions attributed to Big Data could be time-dependent. For example, 40 Gigabytes of data in the 1970's was "Big Data" at that time, whereas the same size of data in 2018 could be processed and stored by possibly a Smartphone.

"Data Analysis is the process of examining data to find facts, relationships, patterns, insights and/or trends" (Erl et al., 2016, p. 21). Data Analysis encompasses the complete data life-cycle, consisting of data collection, data cleansing, data organization, data storage, data analysis and data governance (Erl et al., 2016).

The essence of Big Data analytics is to facilitate the analysis of variety of data formats, be they structured, semi-structured, or unstructured, together for optimum value. Big Data analytics facilitates the combination, integration and analysis of disparate and varied data enblock – to create value irrespective of data sources, type, size or format (Datameer, 2016, p. 1).

Big Data analytics involves the identification, procurement, preparation and analysis of large datasets of raw, unstructured data to extract meaningful information (or insights) that could be used as inputs for pattern identification, enrichment of existing datasets, and performance of large-scale searches (Erl et al., 2016).

As it has already been mentioned, that since the introduction of the World Wide Web, information has never been in shortage. The amount of data that being collected from our cell phones, online shopping stores, and other transactions has opened the door of what is called the Era of Metadata. Though the proponents of Big Data believe that every information is important, there are still stages of filtering and compression that all information collected goes through to ensure the relevant information is reproduced. Data collected in its raw state is of no value to anyone. One cannot perform any analysis of data collected in its raw form from the various sources available. It is imperative that the collected data be subjected to a certain level of processing.

Data cleaning and structuring is a major focus area with Big Data as information is only useful when accurate and relevant. It, therefore, becomes necessary for industries to find the right process to store and mine their **Varied** data of great **Volume** coming through in such high **Velocity** so that none is lost. The very amount, quantity and speed that the information comes in means that it cannot be possibly stored and processed using one server. Instead, it is distributed onto multiple servers which are then collected and compiled before reaching the stage of computation. This access means that analysis is not done on sample data and increases the level of accuracy of analytic reports.

The workflow of transformation on the data collected involves the use of appropriate technologies in acquisition, storage and refinery. Big Data analytics employ techniques used to analyze and acquire intelligence from the datasets. Thus, Big Data analytics can be viewed as a sub-process in the overall process of 'insight extraction' from big data.

In this study, we concentrate on the following five (5) workflows, mapped out under the two main processes of Big Data analytics as seen in Figure 7.1:

- **Data Acquisition and Recording:** we reiterate that data emanates from many sources and applications. Typical sources include textual based data social network feeds, emails, blogs, online forums, survey responses, corporate documents, news and call center logs, etc. There is also videos, audio recordings, multimedia images and much more. It becomes necessary for the process of Big Data to collect all these datasets from the various sources. Although it is the practice of Big Data that all data is relevant, it's still important that the right data is gotten. Some data can be said to be corrupted. Such data are usually data with some missing files, wrong values and invalid data types. Such data cannot be used until the original data is accessed and the necessary corrections made.

  It is stated that "metadata can be added via automation to data from both internal and external data sources to improve the classification and querying. Examples of appended metadata include dataset size and structure, source information, date and time of creation or collection and language-specific information. It is vital that metadata is machine-readable and passed forward along subsequent analysis stages. This helps maintain data provenance throughout the Big Data analytics lifecycle, which helps to establish and preserve data accuracy and quality.

- **Extraction and Cleaning:** Due to the multiple sources from which data comes from in Big Data analytics, the stage of extraction and cleaning is a very important one. The level of extraction to take place would be determined by the type of data source and the type of analytics to

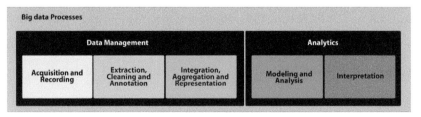

Figure 7.1   Big Data Processes.

Source: www.iti.es, 2017.

be carried out. If the data structure being received is set in the format recognized by the database structure of the Big Data tool being used, the less work is done on the raw data as the dataset can already be processed. Additionally, for the relevance of the output produced, it is expedient that information is well cleansed to remove any possibility of data falsification. In Big Data as most of the information being interpreted comes in unstructured form, the cleaning process removes redundant information, false records, and validates the data as it comes in from the source.

- **Data Integration, Aggregation, and Representation:** Valid data must be void of repetition and redundancy. This, in the ordinary sense, is not so easy to do considering the multiplicity of sources of data in Big Data analytics. All the various fields would be aggregated and the information contained in the multiple fields being the same would be appropriately integrated and in the end, be viewed as one field (Doan, et al., 2012). A form of complex logic is employed to ensure that data integration and aggregation is achieved despite the different form of data structure and language that may be used in the different information set to remove redundancy. E.g. "middle name" and "other name", "last name" and "surname", "sex" and "gender" and the like.
- **Query Processing, Data Modeling, and Analysis:** This stage depends on the data type output or the type of analysis being done; be it a simple or complex analysis. Data analysis can be put into two major categories, namely confirmatory analysis and exploratory analysis.
- **Interpretation:** this becomes the necessary stage after data analysis as all the work done would become absolutely of no value if users cannot understand, interpret or put the results into good use.

There are various tools in Big Data analytics. These tools include;

- **MapReduce:** enables massive scalability across hundreds or thousands of servers in the Hadoop cluster. It basically performs two (2) separate and distinct tasks. The *MAP* is used to convert a dataset into another set of data by breaking down the data and creating tuples (i.e. key/value pairs). The job of *REDUCE* is to take the data from 'map' as input and combines those tuples into smaller set of tuples.

  As an illustration, we consider a simplified dataset containing some social media metrics from Twitter. Using the MapReduce framework which is schematically represented as follows:

$$(input) < k_1, v_1 > \rightarrow map \rightarrow < k2, v2 > \rightarrow combine \rightarrow$$
$$< k2, v2 > \rightarrow reduce \rightarrow < k3, v3 > (output)$$

Let assume, output of the first map yields:

$$< Likes, 60 >$$
$$< Im\,pressions, 980 >$$
$$< Pr\,ofileClicks, 15 >$$
$$< Re\,tweets, 5 >$$

The output of the second map yields:

$$< Likes, 150 >$$
$$< DetailExpands, 12 >$$
$$< TotalEngagement, 29 >$$
$$< Im\,pressions, 1000 >$$
$$< Following, 25 >$$
$$< Re\,tweets, 50 >$$

Upon receipt of the inputs from the map jobs, the reducer sums up the values, which are the event counts for each key; i.e.

$$< Likes, 210 >$$
$$< Retweets, 55 >$$
$$< Following, 25 >$$
$$< Im\,pressions, 1980 >$$
$$< TotalEngagements, 29 >$$
$$< DetailExpands, 12 >$$
$$< Pr\,ofileClicks, 6 >$$

- **Grid Grain:** It provides dynamic methods of computing and storing data such as dynamic clustering, MapReduce processing, distributed closure execution, load balance and fault tolerance, distributed messaging and linear scalability.
- **Hadoop:** The Apache Hadoop library delivers a framework which uses simple programming models for the distributed processing of large dataset across various connected machines. It has been designed to overcome failures and errors at the applications layer, thereby delivering a high accuracy.

7.2.4 Social Media Metrics

Social Media is a form of new media that relies heavily on the participation of users to provide value. They are present with mobile Apps, video, podcasts, e-books, blogs, emails, etc. These media create what is known as findable media object of value. Social media metrics include visitors, visits, page views, and increasingly, events that are so widely used in online marketing and digital business functions; these have become a common currency against which all efforts are judged (Yeboah-Boateng, 2019). Social Media is a class of tools and technologies used to describe media that drives its value from social interactions. It is about formulating and enhancing relationships. This characteristics render social media adept for value creation strategies through customer intimacy.

In systems analysis, the social media relationship is characterized by both one-to-many and many-to-many relations. That is, a social media platform may be able to generate many data sources, such as real-time feeds, raw datasets, historic datasets, cleaned datasets, as well as value-added datasets (Batrinca & Treleaven, 2015). Similarly, different social media platforms could also generate similar datasets as above.

7.3 Methodology

This study employs mathematical or computational approaches to problem-solving with the view to present accurate, precise and concise concepts or discoveries (Bindner & Ericsson, 2011). Applying these principles and adapting the Agile approach, the Big Data analytics coupled with data integration use cases serves as the key algorithmic model in utilizing social media metrics for competitive advantage.

Domestic, transnational and global literature on Big Data, social media metrics and data integration were employed. In this study, we carried out site scraping or web data extraction, which involved collecting online data from social media and other web sites in the form of unstructured text (Batrinca & Treleaven, 2015). We also conducted extensive literature review of key articles, databases, repositories, and authorities, with the view to assessing use cases necessary for competitive advantage when using social media networks and platforms.

Also, cognizance of the agile requirements in respect of benchmarking, we adapted (Lovett, 2011) (Erl et al., 2016) and (Frampton, 2015) social media metrics as the basis for developing the computational equations.

Indeed, the data sources are varied; that is, collecting structured, semi-structured and unstructured datasets. The scope of the datasets gathered is based on the dynamic, interactivity and impression uncertainty characteristics of social media.

In view of the above, and cognizance of the fuzzy nature of the social media metrics, decision-makers perceptions are factored into the treatment in computing for value-added in all cases (McFadzean et al., 2007). In essence, or the key objective is to develop a model that will utilize the experts' opinions to connect the organization's product and services to their customers through customer intimacy programs (Piccoli, 2013).

Theoretically, systems thinking models can be employed to evaluate or assess the utility and reach of social media campaigns, and to explore users' interactions and preferences (Norman, 2009). In assessing social media metrics with evaluation models, (Norman, 2009) recommends the use of dynamic and innovative applications, such as social media.

7.4 Social Media Metric Computations

In this study, we herewith adapt the following baseline metrics from (Lovett, 2011) as the basis for computations, discussions and illustration of use cases which are harnessed for business value creation or competitive advantage.

The adopted social media metrics include:

- Volume
- Engagement
- Influence
- Reach
- Share of Voice
- Impact

"Foundational measures can be used across any social media channel and the individual inputs should be modified to fit each distinct channel. By calculating the measures in the same way, you can create consistency across different platforms and channels" (Lovett, 2011, p. 172). There is the need to establish a baseline first before building the Key Performance Indicators (KPIs) or assigning values to specific metrics (Lovett, 2011). Culled from (Yeboah-Boateng, 2019), Interaction must be measured against a specific marketing initiative. It's evaluated within specific channels or across multiple channels for comparison. It's a composite measure of number of views, unique visitors, shares, conversions surrounding an initiative. The organization can evaluate the percentage of interactions against a benchmark of all visitors. It is given by

$$Interaction = \frac{Conversions}{Activity} \qquad (7.1)$$

It must be noted that Interaction is active, not passive, and it requires sharing, submitting or transacting with the social media channel.

7.4.1 Volume

The volume metric deals with the numeric presence a brand gets across all social media platforms during a reporting period or a specific span of time, which yields statistically relevant data. This metric of volume can be expressed through a variety of social media parameters, including @mentions, shares, links, and impressions. That is, volume is given by:

$$Volume \triangleq \sum_{i=1}^{n} \{Mentions, Shares, Links, Im\,pressions\} \qquad (7.2)$$

Reporting periods are also variable, usually lasting a week, a month, or a quarter. While volume can seem like a simple counting metric, there's more to it than just counting tweets and wall posts. It's important to measure the number of messages about your brand, as well as the number of people talking about your brand, tracking how both of those numbers change over time and being consistent such that you're benchmarking trends with accurate, dependable data (Lovett, 2011).

Taking the most popular social media platforms as use cases to appreciate the impact of this metric, below are needed elements for obtaining good volumetric data.

- **Twitter:** Look at your number of followers and the number of followers for those who *retweeted* your message to determine the monthly potential *reach*. You should track these separately and then compare the month-over-month growth rate of each of these metrics so you can determine where you're seeing the most growth.
- **Facebook:** Track the total number of fans for your brand page. In addition, review the number of friends from those who became fans during a specified period of time or during a promotion and those who commented on or liked your posts to identify the potential monthly Facebook reach. Facebook Insights provides value here.

- **YouTube:** Measure the number of views for videos tied to a promotion or specific period of time, such as monthly, and the total number of subscribers.
- **Blog:** Measure the number of visitors who viewed the posts tied to the promotion or a specific period of time.
- **Email:** Take a look at how many people are on the distribution list and how many actually received the email.

7.4.2 Engagement

This particular metric can be said to be at the core of social media metrics as it shows how many people actually cared enough about what you had to say to result in some kind of action. "Engagement as the total number of unique individual who have liked, clicked, shared or commented on your posts. You can also track the number of people who report or hide content from their newsfeed and mark your posts as spam. You can sort your post reach by organic or paid, fans (people who like your page) or non-fans, and impressions" (Berger, 2018).

Lovett (2011, p. 173) posits that "numerous engagement calculations exist, some of which are as simple as $(time + pageviews)$, and others are extremely complex".

It is given by

$$Engagement = Visits * Time * Comments * Shares \tag{7.3}$$
$$\forall \text{ less engaging topic } \leq 50 \text{ \& more engaging topic } > 50$$

It involves read, converse, comment, and participate.

This metric can be measured by applause rate, average engagement rate, amplification rate etc.

Applause Rate is the number of approval actions (e.g., likes, favorites) a post receives relative to your total number of followers. When a follower likes or favorites one of your posts, she's acknowledging that it's valuable to her. Knowing what percentage of your audience finds value in the things you post. Average Engagement Rate is the number of engagement actions (e.g., likes, shares, comments) a post receives relative to your total number of followers. It's an important metric because higher engagement means your content is resonating with the audience. To prove that, track the engagement rate of every post. If you have a high engagement rate, the actual number of likes and shares and comments is irrelevant (Kallas, 2018).

Amplification Rate is the ratio of shares per post to the number of overall followers. The higher your amplification rate, the more willing your followers are to associate themselves with your brand (Qualmann, 2009). Using our current social media platforms as action-points derived for obtaining an Engagement metric:

- **Twitter:** Quantify the number of times your links were clicked, your message was retweeted, and your hashtag was used and then look at how many people were responsible for the activity. You can also track @replies and direct messages if you can link them to campaign activity.
- **Facebook:** Determine the number of times your links were clicked and your messages were liked or commented on. Then break this down by how many people created this activity. You can also track wall posts and private messages if you can link them to activity that is directly tied to a specific social media campaign.
- **YouTube:** Assess the number of comments on your video, the number of times it was rated, the number of times it was shared and the number of new subscribers.
- **Blog:** Evaluate the number of comments, the number of subscribers generated and finally the number of times the posts were shared and "where" they were shared (i.e., Facebook, Twitter, email, etc.). Measure how many third-party blogs you commented on and the resulting referral traffic to your site.
- **Email:** Calculate how many people opened, clicked and shared your email. Include where the items were shared, similar to the point above. Also, keep track of the number of new subscriptions generated.

7.4.3 Influence

This category gets into a bit of a soft space for measurement. Influence is a subjective metric that relies on your company's perspective for definition (Kelly, 2010). Basically, you want to look at whether the engagement metrics listed above are positive, neutral or negative in sentiment. In other words, did your campaign influence positive vibes toward the brand or did it create bad mojo?

Automation tools such as Twitalyzer, Social Mention, Radian 6 or Scout-Labs make it a little easier, but always do a manual check to validate any sentiment results. Influence is generally displayed as a percentage of positive, neutral and negative sentiment, which is then applied in relation to the engagement metrics and to the metrics for reach where applicable (Kallas, 2018).

A great application for influence is to look at the influence by those who engaged with your brand in the above categories. Do you have a nice mix of big players with large audiences engaging with your brand, as well as the average Joe with a modest following. If not, your influence pendulum may be about to tip over, because it is important that you spend time engaging with both influential users and your average user. Note: many of the automated tools that track sentiment and influence are not free. And many times, you will need a combination of tools to measure all of the different social media channels (Baer, 2010).

Influence is a measure of authority and it is given by:

$$Influence = Volume\ of\ Relevant\ Content *$$
$$Comments * Shares * Reach \qquad (7.4)$$

7.4.4 Reach

Reach measures the spread of a social media conversation. On its own, reach can help you understand the context for your content. How far is your content disseminating and how big is the audience for your message? Reach in social media is a measure to how many people your brand and content are getter in front of. You can think of reach simply as the number of eyes your social media presence is exposed to. Reach is a top-of-funnel metric, but still a metric that every social media marketer should be closely monitoring and continuously working to improve (Kelly, 2010).

Reach is a measure of potential audience size and can be in the post or potential mode. It is given by:

$$Reach = Seed\ Audience * Shared\ Network\ Audience \qquad (7.5)$$

$$Velocity = Reach * Time \qquad (7.6)$$

Post Reach denotes how many people have seen a post since it went live. This type of reach metric is actionable, since it's affected by the timing of post in terms of one's audience being online and the content of your post that is valuable to the audience (Qualmann, 2009) (Lovett, 2011). Potential Reach on the other hand, measures the number of people who could realistically, see a post during a reporting period. If one of your followers shared your post with her network, approximately two to five percent of her followers would factor into the post's potential reach (Shlevner, 2018).

Understanding this metric is important because, as a social marketer, you should always be working to expand your audience. Knowing your potential reach enables you to gauge your progress. And of course, a large audience is good, but reach alone does not tell you everything. Reach becomes very powerful when compared to other engagement metrics. Use reach as the denominator in your social media measurement equations.

Pick important action or engagement numbers like clicks, retweets, or replies (more on this in a second) and divide them by reach to calculate an engagement percentage. Of the possible audience for your campaign, how many people participated? Reach helps contextualize other engagement metrics.

7.4.5 Share of Voice

Here, it considers, how does the conversation about your brand compare to conversations about your competitors? Determine what percentage of the overall conversation about your industry is focused on your brand compared to your main competitors. And learn from your competitors' successes; since so many of these social media conversations are public, you can measure your competitors' impact just as easily as you can measure your own (Lovett, 2011).

$$Share\ of\ Voice = \frac{Brand\ Mentions}{Total\ Mentions\left\{Brand + \sum_{i=1}^{n} Competitor\right\}} \quad (7.7)$$

Consistency and preparation are essential to effective social media measurement. Pick your favorite metrics and start tracking them now. Use the same formulas and tools to calculate these numbers every week or month. Track your numbers over time and pay attention to how they change. If you see anything that looks higher or lower than what you typically expect, investigate it. By measuring – and paying attention to – these five social media metrics, you'll be able to better understand the impact and effectiveness of your social media activity.

7.4.6 Impact

Impact is the ability of a entity to guide the outcome of derived events as measured against specific goals, a.k.a. campaign Return on Investment (ROI). It answers the return of investment questions within social media.

It is a measure of success towards desired outcomes. It is measured in terms of tangible results against expectations set forth when determining business objectives. For example, if the goal of a specific marketing campaign is to acquire a thousand new customers, Impact should be measured as total exposure divided by total new customer acquisitions.

$$Impact = \frac{Total\ Exposure}{Total\ New\ Customers\ Acquisitions} \tag{7.8}$$

Organizations can measure Impact using purely digital metrics and attribution tactics or with a mix of qualitative metrics and anecdotal evidence. It is a critical component of gauging the overall success of organization's social media activity. It is given by:

$$Impact = \frac{Outcomes}{(Interactions + Engagement)} \tag{7.9}$$

Other supporting computations are used, such as the following:

$$Issue\ Resoultion\ Rate = \frac{Total\#Issues\ Resolved\ Satisfactorily}{Total\#Service\ Issues} \tag{7.10}$$

$$Resolution\ Time = \frac{Total\ Inquiry\ Response\ Time}{Total\#Service\ Inquiries} \tag{7.11}$$

$$Satisfaction\ Score = \frac{Customer\ Feedback\ \left\{\sum_{i=1}^{n} Input\right\}}{All\ Customer\ Feedback} \tag{7.12}$$

In order to have competitive advantage using social media initiatives or campaigns, the desired business outcomes must be measurable. These may include metrics or measures such as awareness, comments, posts, engagement and interactions (e.g. sign-ups, navigation, persuasion, conversions, web-hits, purchases, referrals, submissions, etc.) (Sterne, 2010). According to Sterne (2010) irrespective of the metrics or computations adapted, the most important measures are those that drive the "business-critical action" (ibid).

7.5 Social Media Metrics and Use Cases

Big Data has changed and revolutionized the way businesses work together and deliver on their promise. Big Data can be used in every sector and industry. The importance of big data lies in how an organization is using the collected data and not in how much data they have been able to obtain (Laskowski, 2014). There are Big Data solutions that analyze big data much more comfortably than it used to be, which in turn helps the organizations make better and smart business decisions. These benefits include; gaining insights, prediction and decision making, cost-effectiveness, marketing effectiveness. Other examples of applications which Big Data contributes to, are the following, education, healthcare, public sector, transportation, banking industries, etc.

- **Case Example: Banking Sector:** Using this as a Big Data use case and focusing on organizations that handle a large number of financial transactions continue searching for more innovative, practical approaches to fighting fraud. In the traditional fraud detection model, fraud investigators need to work with Business Intelligence (BI) analysts to run complex SQL queries from a bill and claim data, then wait weeks or months to get the results back. This process sometimes causes lengthy delays in legal fraud cases, thus, enormous losses for the business (Juszczak, 2018).

  As a contribution from big data technologies, billions of billing and claim records can be processed and pulled into a search engine so that investigators can analyze individual records by performing automatic searches on a graphical detection to provide automatic results as soon as it recognizes the unfamiliar pattern that matches a previous data. Therefore, with the aid of big data analytics and machine learning, today's fraud prevention systems are highly effective and efficient at detecting criminal activities and preventing false positives (Juszczak, 2018).

  For credit card owners, fraud prevention is one of the most familiar use cases for big data. Even before advanced big data analytics became popular, credit card issuers were using rules-based systems to help them flag potentially fraudulent transactions. Credit card issuers are understandably hesitant about disclosing all the advanced analytic techniques that they use to detect and prevent fraud. However, many credit card firms and other consultants offer technology, advice and services to other firms to help them set systems to criminal transactions. As the issue is very delicate and companies are not willing to cause dissatisfaction with delays in service rendering, payment rejections, etc.

- **Case Example: Social Media Analytics:** The overflow of posts that flow through social media channels like Facebook, Twitter, Instagram and others is one of the most obvious examples of big data. Currently, companies are expected to invigilate what people are saying about them in social media platforms and respond appropriately, and if they don't, they lose clients quickly. Due to that, many industries are investing in tools to help them monitor and analyze social media platforms in real-time. The essence of data exploration was to determine if social media might be used to pinpoint alterations to client request for commodities. The original concept on social media analysis was to distinguish initial social topics and viral events, and link them with products and sales. This ability would enable the business to perceive micro trends and events early sufficient to make knowledgeable ordering, pricing, and product promotions decisions.
- **Case Example: Customer Analysis:** Currently, a lot of initiatives use big data to create a dashboard application that provides an overview of the customer. These dashboards pull together data from a diversity of internal and external sources, evaluate it and project it to customer service, sales and/or marketing personnel in a way that assists them do their jobs. For instance, conceive the type of dashboard a hospital can establish with data about its customers of which in this case study are patients (Laskowski, 2014). Logically, it would include details such as, patients' names, addresses, medical history, family members, lifestyle, genotype etc. The dashboard can obtain vital information from the patient's previous medical records and provide links to present medical conditions, signs and symptoms of possible diagnosis from present lifestyle. All of that information would obviously help prepare the doctor create scheduled check-ups, advice on lifestyle, prescribe drugs that proactively prevents huge disease risks than reactive in order to engage with the patients.

  With big data engineering technologies, you can bring together all of your structured and unstructured data into systems such as Hadoop and analyze all of it as a single dataset, regardless of data type. The analytical results can reveal totally new patterns and insights you never knew existed and aren't even conceivable with traditional analytics.

  Using channels like mobile, social media and e-commerce, customers can access just about any kind of information in seconds. This informs what they should buy, from where and at what price. Based on the information available to them, customers make buying decisions and purchases whenever and wherever it's convenient for them.

- **Case Example: Security Intelligence:** On the subject of scandalous pursuit, associations are also using big data analytics to help them thwart hackers and cyber attackers. Current enterprise IT department generates a gargantuan amount of log data. In addition, cyber threat intelligence data is available from external sources, such as law enforcement or security providers. Big data security solutions vary in sophistication and they are sold under a wide variety of names. Numerous establishments are currently using big data solutions to help them collect and analyze all of this internal and external information in order to prevent, uncover and ease attacks.

7.6 Social Media Metrics Computational and Data Integration Model

The proposed model herein is a Big Data based social media metrics computations and data integration algorithm. The model is presented in Figure 7.2. The proposed model herein is a Big Data based social media metrics computations and data integration algorithm. It depicts the flow processes of data integration of disparate social media metrics received from diverse sources. It also involves the generic business use cases and scenarios, which describe the essence of creating competitive advantage from computing these metrics.

The model is aimed at facilitating digital marketing and corporate transformation using insights from observable computational social media metrics. There are three (3) distinct but dependable phases; i.e. Disparate Sources, Virtual Data Integration, and Value Creation.

The disparate sources consist of entities or actors, be they human actors within the organization, or computer systems, including software systems, and firms. All these sources or entities generate social media metrics in diverse types, formats and at different times and speeds. The datasets generated are fed into the adjourning Virtual Data Integration phase.

Employing the wizard-based data integration techniques, this phase receives the inputs and activates the MapReduce functionality. As per the illustration given earlier on in this treatise, the metrics are mapped out by way of integration, cleaning and preparation. The map outputs are then fed into the Reducer to combine and analyze, employing join and enrichment functions. The resulting outputs are presented as visualized elements or infographics.

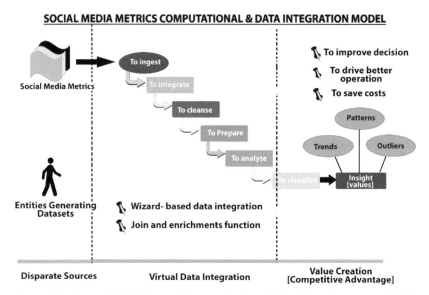

Figure 7.2   Social Media Metrics Computational and Virtual Data Integration Model.

The last phase of value creation basically analyzes the results, make deductions and interprets and/or make inferences to draw insights for business use cases. They often maybe seen as business trends, patterns and even outliers.

It suffices to comment on the perceived value deduced from outliers. Basically, outliers are extreme values that deviate from other observations on the data. They may depict a variability in measurement (c.f. pre-conditions in business use cases), or errors (c.f. post-conditions in business use cases), or a novelty (Santoyo, 2017). This implies that outliers could also give some insights into possible discoveries or findings that may not as yet be obvious. Further research or work may be required to uncover those values. Ultimately, the value created include improved decision-making, to drive better operations and to save costs.

7.7 Conclusion

Generally, Big Data is characteristic of huge datasets that are generated at a fast rate, such as in alive-streaming session, with stakeholders' comments. These datasets are typically unstructured, semi-structured and possibly, structured in formats, with inconsistencies or uncertainties emanating from disparate sources.

Bearing in mind that social media metrics are harnessed for better understanding of today's tech-savvy customers, so that organizations can offer insightful customer experiences and intimacy.

Gartner had predicted that by 2018, 45 percent of the fastest growing organizations will consist of fewer humans than smart machines. In the foregoing, it is clear that organizations can have competitive advantage by utilizing social media metrics collated and integrated employing appropriate Big Data analytics. Though, these innovations abound, it is often the large corporations that are able to afford, whereas the SMEs may not.

The algorithms churning the data are often opaque, and things can go wrong, from the humorous (automated email replies that write "I love you" to a colleague) to the hazardous (a factory robot grabbing a human instead of a machine part). Today's data scientists are developing next generation approaches to make machine learning more approachable and that's the reason for contributing in deploying a cloud-based machine learning. The cloud-based machine learning application has been built in way that an organization doesn't need to have data scientists or know the modules before it can access it. This application with the provided data keyed into the system can analyze data of different varieties anonymous data within the shortest possible time. The contribution of this application automates some of the work of data scientists due to the lack of skilled personnel's in such administration.

This study has demonstrated that social media metrics could be harnessed as business motivation and driver towards Big Data analytics amongst SMEs in developing economies, for example. Corporate executives ought to appreciate that Big Data isn't just about emerging technologies but also about how the insights derived from these datasets can be harnessed to improve decision-making, to drive better operations, and to cut costs (Erl et al., 2016).

Business delivers value to customers through the execution of their business processes. In essence the business objectives are use cases with the executives, customers and partners as the actors. For example, employees as well as customers are co-creators of the product brands and corporate reputation; such as using social media metrics to provide unique and differentiating value propositions. Furthermore, businesses leverage on innovative cloud computing capabilities and resources to provision Big Data solutions using external datasets. That is, leveraging on social media metrics, businesses can rent and/or provision scalable processing facilities and available huge storage opportunities to create value (Erl et al., 2016).

Future studies would dwell on collecting actual social media metrics datasets and to perform an experiment to ascertain the veracity or otherwise of the se computations.

References

Agarwal and Dhar, (2014). Critical Analysis of Big Data Challenges and Analytical Methods, 70, pp. 263–286.

Akred, J. (2014). *What is Big Data?* Silicon Valley Data Science. Available: http://datascience.berkeley.edu/what-is-big-data/. [Accessed 14 December 2018].

Batrinca, B., and Treleaven, P. C. (2015). Social Media Analytics: A Survey of Techniques, tools and platforms. *Ai and Soc*, 30(1), 89–116.

Bindner, D. and Ericsson, M. (2011). *A Student's Guide to the Study, Practice and Tools of Modern Mathematics*, CRC Press.

Datameer, (2016). *Top Five High-Impact Use Cases for Big Data*, Datameer.

Doan, A., Halevy, A., and Ives, Z. (2012). *Principles of data integration.* Elsevier.

Dobre, C. and Xhafa, F. (2014). Intelligent Services for Big Data, *Future Generation Computer Systems,* vol. 37, pp. 267–281.

Erl, T., Khattak, W., and Buhler, P. (2016). *Big Data Fundamentals: Concepts, Drivers and Techniques,* ServiceTech Press, Prentice Hall.

Frampton, M. (2015). *Big Data Made Easy: A Working Guide to the Complete Hadoop Toolset*, Apress.

Hahn, J. (2017). *The Internet of Things: Mobile Technology and Location Services in Libraries*. American Library Association.

Juszczak, R. (2018). Mobile Banking Case Study.

Kallas, P. (2018). 48 Social Media KPIs.

Kelly, N. (2010). *8 Social Media Metrics You.* Available: www.socialmediaex aminer.com/

Laskowski, N. (2014). The Big Data Market Rises and Must Converge – Just not Yet, TechTarget.

Lovett, J. (2011). *Social Media Metrics Secrets*, Wiley Publishing, Inc.

McFadzean, E., Ezingeard, J. N., and Birchall, D. (2007). Perception of Risk & the Strategic Impact of existing IT on Information Security Strategy at the Board level, 30(5), pp. 622–660.

Norma, C. D. (2009). Health Promotion as a Systems Science and Practice, *Journal of Evaluation and Clinical Practice*, 15, pp. 868–872.

Piccoli, G. (2013). *Information Systems for Managers: Text and Cases*, 2nd ed., John Wiley & Sons, Inc.

Planet, B. (2012). Structured vs Unstructured Data. Available: www.brightpla net.com/06/structured-vs-unstructured-data/. [Accessed 14 December 2018].

Poulson, B. (2014). Big Data Foundations, Techniques and Concepts.

Qualmann, E. (2009). *Socialnomics: How Social Media Transfoms The Way We Live and Do Business*, John Wiley & Sons, Inc.

Randolph. B. (2017). The Beginner's Guide to Structured Data for SEO: A Two-Part Series.

Raza, M. (2018). Introduction to Database Management Systems (DBMS), Available: www.bmc.com/blog/dbms-database-management-systems/. [Accessed 14 December 2018].

Santoyo, S. (2017). A Brief Overview of Outliers Detection Techniques, Towards Data Science. Available: www.towardsdatascience.com/a-brief-overview-of-outlier-detection-techniques. [Accessed 15 December 2018].

SAS Institute. (2007). SAS Data Integration Studio 3.4: *User's Guide*, SAS Institute.

Shacklett, M. (2017). *UnStructured Data: A Cheat Sheet*, TechRepublic.

Shlevner, E. (2018). *Social Media That Matter*, Available online: www.hootsuite.com/author/eddie-shleyner.

Sterne, J. (2010). *Social Media Metrics: How to Measure and Optimze Your Marketing Investment*, Hoboken, NJ: John Wiley & Sons, Inc.

Swafford, S. (2006). Use Cases and Their Importance. Available: www.aspalliance.com/765+Use_Cases_and_Their_Importance.all. [Access ed 14 December 2018].

Yeboah-Boateng, E. O. (2013a). *Cyber-Security Challenges with SMEs in Developing Economies: Issues of Confidentiality, Integrity and Availablity (CIA),* center for Communications, Media and Information technologies (CMI), Aalborg University, Copengahen.

Yeboah-Boateng, E. O. (2017). Cyber-Security Concerns with Cloud Computing: Business Value Creation and Performance Perspectives, in Resource Management & Efficiency in Cloud Computing Environment, IGI Global Publishers, pp. 106–137.

Yeboah-Boateng, E. O. (2019). *Generating Big Data: Leveraging on New Media for Value Creation*, in Gyamfi, A. and Williams, I. (eds). Big Data and Knowledge Sharing in Virtual Organizations, Vols. Advances in Knowledge Acquisition, Transfer and Management (AKATM), IGI Global, pp. 136–161.

8

Intrusion Detection and Prevention System for Wireless Sensor Network Using Machine Learning: A Comprehensive Survey and Discussion

Pankaj R. Chandre, Parikshit N. Mahalle, Geetanjali R. Shinde and Prashant S. Dhotre

Department of Computer Engineering,
Savitribai Phule Pune University, Pune, India
pankajchandre30@gmail.com, aalborg.pnm@gmail.com,
gr83gita@gmail.com, prashantsdhotre@gmail.com

Network security is a protected method to access files, directories, and data in a network against hackers, misuse and unauthorized changes to the system. Network security is an activity, which is intended to safeguard the usability and integrity of the network as well as the data. The network security encompasses a mixture of computer code and hardware used to target a group of threats and it stops them from stepping into the system. Intrusion detection and prevention is one of the main research problems in the field of network security. The aim is to spot unauthorized access or attacks to secure internal networks. In a literature survey, the intrusion detection is described and presented by numerous techniques. It is further discussed why the firewall generally isn't ready to observe attacks completely which can result in knowledge loss. Using machine learning techniques, an intrusion prevention system can be enforced to stop misuse of the information and network. Machine learning has significances like knowledge analysis, devel-oping predicting models and to classify bound set of knowledge in numerous forms. Intrusion is one amongst the vital problems, and intrusion detection has already reached to the saturation. This chapter presents a comprehensive survey on intrusion detection and prevention systems for Wireless Sensor Networks (WSN) with the aim of identifying and analysing the gap that exists in the systems. Further, this chapter also proposes a method for IDP in WSNs using machine learning technique(s) that will sufficiently bridge the gap/s identified. Effectively, this chapter suggests a need of an efficient solution for proactive intrusion prevention towards wireless sensor networks.

8.1 Introduction

The Internet has become an inevitable tool and plays an important role in our day-to-day lives. It helps the people in several areas, such as business, education, entertainment and many more. In general, the Internet has been used as an essential component of business models. This requires the appropriate information security measures in the Internet as the media needs to be carefully concerned. Detection of intrusion is one of the major research problems in communication networks such as business and personal networks. By using the Internet, there is a significant risk of network attack, at the same time there are various systems designed to block the Internet-based attacks (Tsai et al., 2009).

Intrusion detection systems can be categories into two types based upon the detection approaches/mechanisms. They are anomaly-based detection and Signature-based detection. An anomaly-based intrusion detection system monitors the activities of the network and the host system and by result classifies it as either normal or abnormal activity (Das & Nehe, 2017). Again, the classification is based on some rules instead of signatures and patterns. Such type of systems work exactly opposite to signature-based detection systems in which attacks are detected by considering the signatures, which are created previously. On the other hand, a Signature-based detection system can identify attacks by looking for specific patterns such as byte sequences used for transmission in network traffic or malicious public course of instruction used by some malware. A signature-based detection system is able to detect known attack very quickly, but it is difficult to find out new attacks by using same system (Moon et al., 2014).

An Intrusion prevention system is sometimes called a device or software that has the capabilities of an intrusion detection system and apart from the these capabilities the intrusion prevention system can stop the possible attacks. Currently, the Intrusion Detection interference System (IDS) plays an important role in stopping the quantity of cybercrimes. When an intrusion takes place, the attempt might be intentional or unauthorized to access data, to modify data, or to make the system unusable (see Intrusion detection system in reference). In terms of computer security, IDS help in detecting and preventing unauthorized use (destruction/interruption) of resources such as computer/data (i.e., hardware or software). In computer security, intruders are prevented from using the computers to their benefit. Let us consider some examples of system security such as firewalls, anti-viruses, etc. (see SANS Institute in reference The four main areas of concern in computer security are availability of data at the right time, authentication of resources, confidentiality of documents and integrity of all the specified data. Normally, IDS is placed inline, it actively analyses and takes automated actions on all traffic flows that enter the network (Das & Nehe, 2017).

A Wireless Sensor Network (WSN) may be a part of a system that Senses, collects and analyses environmental information (temperature, humidity, sound, and image), and provides responses to users. A wireless sensing element network consists of multiple sensing element nodes and one or more sink nodes. The primary features of WSNs are limited resources, minimum memory, and low bandwidth. Because of the above features, the wireless sensor networks are vulnerable to security attacks (Ashoor & Gore, 2001). Wormhole, Sybil and selective forwarding are the possible attacks at network layer. In forwarding attacks, compromised nodes in the network maliciously drop some content or all content of reports that are proceed through the routing paths that contain them (Athanasios, 2011). In wormhole and Sybil attacks, attackers try to modify routing paths so that compromised nodes are included in multiple routing paths (Hellbusch, 2014).

8.1.1 Overview of the Problem

Intrusion Prevention System (IPS) is employed not only to observe malicious activity, but as the name suggests it executes a timely response to prevent a direct threat to your network. Now-a-days, Wireless Sensor Networks are widely used in military applications and by its applications; it is extended to healthcare, industrial environments and many more. Despite the use of WSN has its limitations such as limited power supply, minimum bandwidth and limited energy. The application of the same techniques used to secure the traditional networks may not prove effective for WSN due to the above said limitations.

Therefore, to increase the overall security of WSNs, new ideas and new approaches are required. In general, intrusion prevention is the primary issue in WSNs and intrusion detection already reached to saturation. Thus, we need an efficient solution for proactive intrusion prevention towards WSNs.

8.1.2 How this Problem Affects India or Pune

In the next couple of years, IPS will become the top seed for intrusion systems and it has already started replacing the IDS. The only effective way to know how an IPS appliance will affect your network is to put it in line and see what happens.

8.1.3 Why There is the Need for the Solutions You are Providing

Amongst the many reasons that exist for the need of intrusion prevention systems, protection from denial of service attacks and protection from many critical exposures are the most important ones. The capabilities of IPSs are already put to use by large organizations and in the near future we will more than likely see private home users utilizing a variation of IPS. More discussions have been made on IDS, however it is not effective enough as it deals with intrusions that have already taken effect on the system.

The need for an IPS can be explained well with the help of an example from the banking domain (COSMOS BANK), where security is a crucial issue. The banking system has been exposed to an oversized variety of cyber-attacks on their information privacy and security, like frauds with on-line payments, ATM machines, electronic cards, web banking transactions, etc. Cyber-attacks routed towards financial organizations are fourfold in comparison to those targeted towards other organizations/industries. The motive behind these intrusions is to achieve confidential information or to steal cash from banks despite stringent security measures being enforced to the transactions. However, the number of attacks done by breaching the protection has doubled thereby inflicting great loss to the banks and in turn the associated people. Hence, there is a greater need to implement counter measures that will help in guarding the valuable assets of the bank against such attacks. Implementing security into the above said system includes incorporating security into the associated network, computer systems and all the possible components that could be potential preys to the intruders.

8.1.4 Why these Interventions/Intrusions are Important

In order to explain the importance of addressing such intrusions, we present a much-simplified example. Consider a residence which is occupied by the members of the family who are engaged in their work place from 11:00 am to 5:00 pm. The house is empty and left without supervision during the period. Intruders can take advantage of this time period and steal valuable objects from the house. Hence it is important to deal with such intrusions as it might lead to adverse effects.

8.2 Machine Learning

Machine learning belongs to the category of data analysis, which automates analytical model building. In other words, it is a branch of data analytics that supports the notion that a system can learn from information, establish patterns and create choices with the least human intervention (Chanaky et al., 2017). Following are the various techniques in Machine learning (Singh & Nene, 2013): Supervised learning, unsupervised learning, reinforcement learning and Semi-supervised learning.

8.2.1 Supervised Learning

It's a sort of machine learning technique that trains a model on famous input and output information to predict future outputs. Supervised learning is to be used if there is some prior known information on the dataset for which a prediction on category is to be made. In supervised learning, develop a prophetical model regression and classification techniques are used (Kim & Aminanto, 2017). Classification techniques are employed to predict separate responses.

Here, we contemplate an example, within which we tend to square measure planning to check the received email in terms of spam or real. In this, classification models square measure to classify input file. We will use classification techniques, once our information is well categorized or once our information is correctly labelled intervention (Chanaky et al., 2017). Regression techniques are used to predict continuous responses.

8.2.2 Unsupervised Learning

Unsupervised learning is a class of machine learning that is employed to seek the patterns that square measure hidden within the input file. Subsequently, with the assistance of the same we draw interferences from the datasets that consists of input file with untagged responses (Kim & Aminanto, 2017). Finally, the cluster is employed to perform information analysis to look out for groups in the given information or any hidden patterns.

8.2.3 Reinforcement Learning

Reinforcement learning is a category in machine learning wherein the simplest actions are supported by a reward, or penalization. Basically, in reinforcement learning, we tend contemplate three ideas like state, effort, and compensation.

State: State deals with the current situation. For example, for a robot state is nothing but the position of its two legs to learn walk.

- Action: Action is nothing, but the user can do in each state. For example, the robot can take action like distance to cover in between steps.
- Reward: Reward is obtained upon performing some actions.

8.2.4 Semi-Supervised Learning

In general, in semi-supervised learning, unlabeled information is used in order to achieve understanding of the population structure. Considering a straightforward example, a glance at the Kaggle state farms dataset with 60000 (considering an oversized population) pictures, their square measures will solely include 15000 pictures which forms the coaching information set. This suggests that we should always realize the simplest way for analysis of giant unlabeled information. Thus, we need both intrusion detection and intrusion prevention schemes in a WSN to secure our network.

8.3 Motivation

Due to the tremendous increasing demands from economic aspects and from the users, ubiquitous computing is moving towards a dynamic and distributed next-generation network. In the sequel, wireless computer and sensor networks has become an integral part on the day-to-day lives of users.

Here, we would like to elaborate on the well-known cyber-attack wherein Pune's Cosmos Bank loses 94 Cr. When a transaction takes place using a debit card, an ATM machine connects to the switch which has default password and then in turn connects to the banking server. The switch is then manually authenticated to the banking server. Now, an attacker had deployed some malware on the switch leading it to be considered as a genuine switch effectively routing all transactions through this replicated/compromised switch. By employing a parallel proxy switch and cloned debit cards, the hackers have managed to self-approve the transactions and withdraw the cash. Now a days, malware support spoofing and man-in-the middle attack are prevalent. Attackers also use the conception of phishing to introduce a malware into the industry. This attack happens owing to the switch having default positive identification.

Let us consider a simple example of a cell phone, where security is an obvious issue. Now a days, cell phones have become an important part of everyone's lives. User's entire life is documented on the cell phone because it contains user's private pictures, bank details, personal emails, call logs, contacts and other personal information. So, with the rapidly growing importance of cell phones the device has become a valuable item in terms of personal data stored in it and for the same reason, the cell phone theft has become a serious problem by which we can lose our personal data.

Due to *"anything anytime anywhere"* means of computing, use of wireless sensor network has increased rapidly. Looking at the various threats and several attacks in a wireless network, security is a prime concern. Intrusion is one of the critical issues, and intrusion detection has already reached to its saturation. Thus, we need an efficient solution for proactive intrusion prevention towards WSNs.

8.4 Overview of IDS and IPS

This section deals with the overview of intrusion detection system briefing us on the following points.

- What an intrusion detection system is?
- Why is the need for an intrusion detection system and how it works?

8.4.1 An IDS Overview

This section provides a short overview of intrusion detection system.

- **What is an IDS?**

An IDS is a device or software that is used to monitors computer systems, and then it looks for signs of intrusion concerning unauthorized users or authorized users who try overstepping their permissions (Yousufi et al., 2017). An intrusion detection system used to monitor a variety of data from computer systems then performs analysis on the same information in different ways. The first way is that it compares information with the available large number of databases of attack signatures, then each reflecting an attempt to bypass security protections. The second way is that it looks for the problems which are related to authorizing a user who tries to overstep their bounds (Yousufi et al., 2017).

- **Why do we need IDS?**

The firewall is employed to produce security that is like a security fence around a land and the security guard who focuses his/her attention at the front gate solely.

Anything that passes through the gate has to be screened by the security guard, however he/she has no idea about the safety measures/happenings within the compound. Thus naturally, we can say intrusion detection systems are like multi-sensor video observation with stealer alarm systems (Yousufi et al., 2017) . With the assistance of the system, all information is collected centrally, after which analysis is performed on the collected information. If any suspicious activity is sensed, then some actions are to be taken to stop that activity.

- **How does IDS work?**

An intrusion detection system collects and analyses data from within a computer or network to identify misuse, unauthorized access, and some possible violations. Sometimes, an intrusion detection system can be considered as a packet sniffer which intercepts packets travels along various communications medium (Mahurkar & Athawale, 2016). An analysis is performed on all the captured packets. The main aims of intrusion detection system are not only intrusion prevention but also to give alerts to administrators when the attack is going on.

Figure 8.1 shows the flow of IDS. The working of an intrusion detection system is given below (see GB Hackers in reference):

1. The Intrusion detection system has sensors which are used to detect signatures.

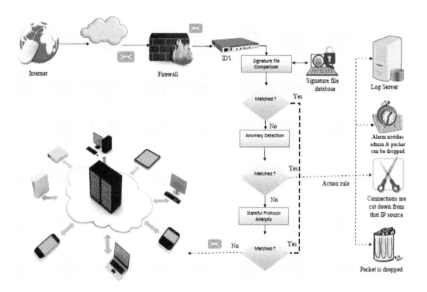

Figure 8.1 Flow of an IDS.

2. Some advanced intrusion detection systems incorporate a behavioral activity monitoring system to detect malicious behaviors. If there is a match with the signature, then the system will inform the occurrence of an attack.
3. If there is a signature match, then successive steps will be done. Otherwise, the packets are dropped, and the alarm is declined.
4. Once there is a signature match, the control is passed on to an anomaly detection system to decide whether the received packet or request matches the signature.
5. After receiving packets from an anomaly stage, the operation of stateful protocol analysis is finished. After that with the help of a switch, packets are given on the network. If again there is any mismatch, the connection with a source IP address is cut down, and packets are dropped, and alarms will be raised to notify an administrator (see GB Hackers in reference).

- **What are the types IDS?**

This section elaborates the different types of intrusion detection system are as follows:

1. **Network-based IDS:** Network-based intrusion detection system collects information from the network itself rather than receiving it from individual hosts. By inspecting the contents and the header information of all packets, a system will check for attacks or some misuse. The network sensors are equipped with attack signatures associated rules. The sensor area units compare these signatures with the captured traffic. The attack recognition module uses 3 ways to acknowledge attack signatures (Shoaib et al., 2017).

   - pattern, expression or bytecode matching
   - frequency or threshold crossing
   - statistical anomaly detection

2. **Host-based IDS:** Host-based intrusion detection systems deal with a group of data from a system or a host. A system is dedicated to behave as a vulnerable system thereby enabling it to attract attacks. The sensors area unit monitors these vulnerable hosts and collects and records information regarding every suspected event happening in such hosts. By mistreatment audit trails as associate operative systems mechanism, this information is recorded. Host-based sensors area unit keep track of the behavior of an individual user (Patil et al., 2017).

8.4.2 IPS Overview

This section deals with the overview of intrusion prevention system like what is intrusion prevention system, why we need intrusion prevention system, how it works.

- **What is IPS?**

An Intrusion Prevention System has some distinctive characteristics which, sets it aside from other available security solutions. Intrusion Prevention System is employed to monitor network traffic for signs of a potential attack. Once it detects some peculiar activity (in the shape of attacks), it takes some action to prevent that attack. This action is within the approach of dropping malicious packets, by obstruction network traffic or by resetting the network connections. An Intrusion Prevention System, in addition to the above said actions, sends associate conscious to the protection directors concerning an equivalent potential malicious activity (attack). Considering the state-of-the-art mechanisms, most of the intrusion hindrance strategies depend heavily on humans to perform analysis of knowledge and classify it into intrusive and non-intrusive networks. Once the information to reason becomes extremely large, computing strategies like processor intelligence and its constituent are essential to use as humans have their shortcomings (IPS White paper, 2013).

- **Why do we need IPS?**

An IPS is the essential things for many organizations in order to detect and stop attacks.

1. For identification and interference of attacks that the competitor security controls cannot do.
2. For customization of detection capabilities to prevent malicious activities that are solely of concern to one organization; and
3. To minimize the quantity of network traffic.

- **How does IPS works?**

IPS is placed in between a firewall and the remainder of the network and is used to prevent the malicious activity from attending to the remainder of World Wide Web. If your IPS may be a watchman, United Nations agency will forestall attackers from getting into your network. Once a noted event is detected, then that activity is stopped.

Associate Intrusion interference System works in in-line mode. It contains a device that is found directly within the actual network traffic path, which is successively employed to deeply inspect all the network traffic that passes through it (see GB Hackers in reference).

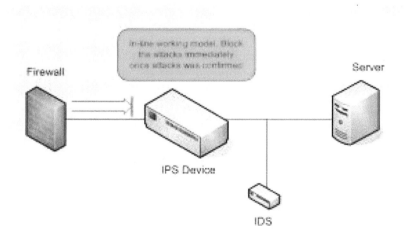

Figure 8.2   Shows the working of IPS.
Source: See Intrusion detection System in Reference.

- **What are the types of IPS?**

Different types of intrusion prevention system are as follows:

1. **Network-based IPS:** Network-based intrusion detection system is employed to observe and analyse the network traffic to shield a system from network-based threats. A network-based intrusion detection system reads all incoming packets and searches it for any suspicious patterns. If threats are discovered, the system will act accordingly by notifying directors or prevent the supply science address from accessing the network.

2. **Wireless-based IPS:** A wireless-based interference system may be a dedicated security device or integrated computer code application that monitors a wireless local area network's spectrum for scalawag access points and alternative wireless threats. A wireless-based interference system performs its task by scrutinizing the mackintosh addresses of all wireless access points on a network against the noted signatures of pre-authorized, noted wireless access points and sends associate responsive to associate administrator once any form of discrepancy is found (Stiawan et al., 2011).

3. **Network behaviour analysis-based IPS**: Network behaviour analysis-based IPS may be viewed as a system to enhance the safety of a non-public network by observing the traffic for differentiating uncommon actions from traditional operation. Ancient IPS solutions defend a network's perimeter by mistreatment packet examination, signature detection, and period interference found (Stiawan et al., 2011).

4. **Host-based IPS:** A host-based IPS, may be a system or a program that is used to shield systems containing crucial information against viruses and equivalent web malware. Ranging from the network layer to the application layer of the OSI model, the host-based intrusion interference system protects the system from well known and unknown malicious attacks found (Stiawan et al., 2011).

8.4.3 Difference between IDS and IPS

The main difference between IDS and IPS is the form of action they take when an attack is detected by network scanning and port scanning. The differences between IDS and IPS are categorized under four objectives as follows (Ashoor & Gore, 2001).

- **Network stability and performance:** Generally, an intrusion prevention system is deployed out of the band in a network and is used to pass all network traffic to a given system, via intermediate devices and the processing capacity of IDS is made to match the average network load. There is a variation in the latency between the captured and reported packets which ranges from a few seconds to minutes. However, this latency depends upon the human response time. Still, IDS is a type of logging device which requires more memory buffers to deal with the traffic bursts and average network loads (Ashoor & Gore, 2001). An IPS is deployed in-line in a network to pass the traffic through intermediate devices, which are connected.
- **Accuracy – False Positives:**In IDS and IPS, there are three rules to calculate accuracy concerning false positives:
  1. An IDS has minimum false positives, and an IPD has no false positives.
  2. An IDS false positive is used to give alerts on an intrusion that may or may not be true. An IPS false positive is used to block legitimate traffic.
  3. In both IDS and IPS, the anomaly filters are not used to block the traffic.
- **Accuracy – False Negatives:** The accuracy of false negatives is found out by measuring the missed attacks. The goal of such a system is typically based on coverage of high priority attacks (Ashoor & Gore, 2001).
- **Datalog analysis:**An IDS and IPS system is used to give comprehensive logging an d information collection capabilities with actionable alerts. The data collected from these systems and sensors can be used for event correlation and network forensics in a post-attack scenario. This type of data is very critical to perform analysis during and after attacks (Ashoor & Gore, 2001).

Table 8.1 shows the difference between intrusion detection and prevention system. For differentiation, some parameters considered here are, the concept, data used, and placement, how it works, configuration mode, action, traffic pattern and related technologies.

8.4.4 Threats

Threats are potential causes for vulnerabilities that can be turned into attacks on systems, networks and all devices. Because of the dangers, individual's

Table 8.1   Difference between IDS and IPS

Parameter	Intrusion Detection System	Intrusion Prevention System
Concept	It is a device or application program that analyses complete packets, both header, and payload, looking for known events. When a known fact is detected a log message is generated detailing the incident.	It is a device or application program that analyses complete packets, both header, and payload, looking for known events. When a known fact is detected the packet is rejected.
Data used	It works on Header only.	It works on Header and payload.
Placement	It is used offline.	It is used inline.
How does it work?	Observes and detects traffic, looks for the pattern or signature and then generates an alert.	Observes and detects traffic, looks for the pattern or signature and prevents if detected.
Configuration mode	End-host.	Layer 2.
Action	Alerts the system.	Prevents malicious activity.
Traffic pattern	Analysed.	Analysed.
Related terminologies	• Anomaly-based detection • Signature-based detection • Zero-day attack • Blocking the attack • Raising alarm	• Anomaly-based detection • Signature-based detection • Zero-day attack • Preventing the attack

computers or business computers are at risk. In order to mask/rectify these vulnerabilities a fix is required so that the attackers cannot gain access to the system and cause damage. Threats can include everything from viruses, Trojans, backdoors to outright attacks from hackers.

Basically, threats can be divided into two types (Gheorghe et al., 2010). These types are as follows:

- Soft threats
- Hard threats

Based on these categories, warnings can be classified as:

- **Physical Threats:** Physical threats may directly destroy the computer systems hardware and infrastructure. Some of the internal threats include power supply and humidity. Earthquakes or any kind of natural disaster is an external threat. Most of the human involved risks are either intentional or accidental.

- **Non-physical Threats:** The non-physical threats are used to target the data and other software which are installed on a computer system. Some of the non-physical risks include virus, spyware, malware, adware, Trojans, phishing, DOS attacks, DDOS attacks, unauthorized access to systems or networks, etc. (Yi et al., 2012).

Table 8.2 shows the wireless sensor networks layer-wise threats and related countermeasures. Before preventing any type of attack, it is necessary to find out risks related to that attack.

Table 8.2   Wireless Sensor Network layer-wise threats and countermeasures

Layer	Threats	Countermeasures
Application layer	Malicious Node	Integrity Protection
Transport layer	Denial of service attack	Manage connection request.
Network layer	Wormhole	Physical monitoring of Field devices and regular tracking of network using Source Routing. The monitoring system may use Packet Leach techniques.
	Selective forwarding	Regular network monitoring using Source Routing.
	DOS	Protection of network-specific data like Network ID etc. Physical security and inspection of a network.
	Sybil	Resetting of devices and changing of session keys.
	Traffic analysis	Sending a dummy packet in quiet hours; and regular monitoring WSN network.
	Eavesdropping	Session keys protect NPDU from Eavesdroppers.
Data Link layer	Collision	CRC and time diversity.
	Exhaustion	Protection of Network ID and other information that is required for joining device.
	Spoofing	Use different path for re-sending the message.

(Continued)

Table 8.2   *(Continued)*

Layer	Threats	Countermeasures
	Sybil	Regularly changing a key
	De-synchronization	Using different neighbours for time synchronization.
	Traffic analysis	Sending of a dummy packet in quiet hours; and regular monitoring WSN network.
	Eavesdropping	Key protects DLPDU from Eavesdropper.
Physical Layer	Interference	Channel is hoping and blacklisting.
	Jamming	Channel hoping and blacklisting.
	Sybil	Physical protection of devices.
	Tampering	Protection and changing of a key.

Source: (Hellbusch, 2014).

8.4.5 How machine learning plays an essential role in IDS and IPS

Machine learning plays a critical role in our day-to-day lives. For example, when we load an image on social media, we might be prompted to tag other people to that image. This is called as image recognition which is a capability provided by machine learning through which the computer can learn to identify facial features. From an intrusion detection perspective, security analysts can apply a concept of machine learning to differentiate normal and malicious traffic (see Intrusion detection system in reference). A computer can learn to recognize an object, such as a bike. The laptop can extract features from the motorcycle such as its colour. Further, from colour, we can model it. Security analysts will use the thought of machine learning to create a helpful intrusion detection capability. So, there is a requirement for machine learning with an intrusion detection system. An intrusion detection system is employed to observe the network traffic by searching for suspicious activity, that may represent an attack or unauthorized access. Older strategies were designed to notice famous attacks however could not determine unknown threats.

The most discreetly famous risks supported outlined rules or activity analysis through baselining the network. A sophisticated attacker will bypass these techniques, thereby necessitating intelligent intrusion detection in mordern days. So, researchers are attempting to use machine learning techniques to the current space of cyber security.

We can apply machine learning to improve intrusion detection system by considering the following parameters:

- Network traffic analysis
- Future extraction
- Creating a useful dataset
- Selecting and classifying features

In supervised learning, the system works with labelled events which occur in the network. This type of learning is similar to signature-based IDS with a slight difference. A different attack event which is created by network flow data is used in the training phase of supervised learning. In signature-based IDS/IPS attack signatures are used but in anomaly-based IDS/IPS network flow is used. There are many supervised learning techniques like support vector machine, a Bayesian network, artificial neural network, decision tree and k-nearest neighbour. The advantage of this is that they can recognize well known malicious activities with high accuracy and low false alarm rate.

In unsupervised learning approaches, the data set doesn't consist of any class information. In these approaches, we can't change the quality of the profile within a short time, and malicious activity causes an abnormal change in network traffic.

8.5 Related Work and Gap Analysis

Similar work in the field of discussion and gap analysis of the discussed methods adopted are presented in this section. A discussion on the existing literature related to the topic is presented below.

In Das & Nehe (2017) the authors have proposed periodic models, which analyse and handle large datasets in addition to subtle information and can deliver it quicker with correct results. Deep learning is considered as a sub-field of artificial intelligence which possesses all of the attributes. Intrusion preventions integrate methods that help to prevent intrusive and non-intrusive data packets. Most of the intrusion prevention methods depend on humans to analyse the data and classify it into intrusive and non-intrusive networks. When the data that we must compute is in a significant amount, computing methods such as Computational intelligence and its constituent are essential to use as humans have their shortcomings. Computational intelligence and deep learning are widely used for detection of intrusion, which in turn helps to prevent the same (Das & Nehe, 2017). Pattern recognition uses algorithms to handle massive datasets.

In Singh & Nene (2013), the authors have expressed that currently, on a daily basis, machine learning for intrusion detection has received an excessive amount of attention within the field of Artificial Intelligence (AI). In intrusion detection formula, a big quantity of information should be analysed to construct new detection rules for increasing range of attacks in an exceedingly normal network. By considering the advanced properties of intrusion detection formula, attacker's behaviours square measure has improved in terms of detection speed and detection accuracy. By analysing the network dataset and improved performance of detection accuracy, intrusion detection has become extremely vital within the field of deep learning (Singh & Nene, 2013). The major advantage of a system supported by deep learning is its ability to observe or categorize features while not taking any feedback from the surroundings, thereby overcoming the disadvantage that exists in normal systems with respect to the overhead involved in the training of the system.

It's a difficult task to decide on alternative formulas for setting the intrusion bar. Victimization support vector methodology, overcomes the static program directions by creating data-driven predictions or choices, through building a model from sample inputs. Some forms of attack, such as virus can be tackled by installing anti-virus, however it slows down the system, as it uses heaps of memory.

In Kim & Aminanto (2017) the authors have investigated various significant algorithms in intrusion detection. They believe that adopting nature's policy, can improve the current methods. They started by observing the behaviour of ants and constructed the Ant Clustering Algorithm (ACA) (Kim & Aminanto, 2017). However, they need other methods for improving the performance of the IDS. We believe that ACA is still limited to distinguishing benign and attack instances. Therefore, they have adopted deep learning, which is the advanced part of a neural network. Incorporating deep learning methods as a real-time classifier will be a challenging task. Majority of previous works that have leveraged deep learning methods in their IDS environment have limited the performance to feature extraction or reducing feature dimensionalities only. The authors have concluded saying that, detection is not only the measured by the level of security offered but also needs to be extended towards intrusion prevention. So, in future, there is a need to design a system that will be useful for intrusion detection as well as intrusion prevention.

The proposed system performs only intrusion detection and is unable to perform intrusion prevention. So, there is a need to design an IPS, which can detect and prevent zero-day attacks with large detection rate and low false alarm rate. Improving the unsupervised approach is necessary since huge labelled data are difficult to obtain.

In Patil et al. (2017) the authors have proposed neat Cooperative intrusion detection and hindrance design for se nsible Grid ecosystems. The Collaborative Smart Intrusion Detection Prevention System (CSIDPS) planned utilizes three advanced elements like involuntary manager, data manager, and mathematical logic risk manager that were deemed essential for associate degree Smart Grid (SG). By exploiting this, the simulated system portrays a more robust detection rate with low false positive alarms. This technique, with a mixture of advanced cooperative sensible soft computing and involuntary computing elements, was a unique approach in associate degree IDPS setting. This CSIDPS framework overcame the recent challenges in detecting unknown vulnerabilities with lower false positives and false harmful alarms that resulted in higher detection accuracy than the normal IDPSs (Patil et al., 2017). Future enhancements to CSIDPS would be to incorporate a comprehensive digital forensics element as a stratagem since SG could be a vital infrastructure that should be adequately protected, and cyber culprits prosecuted with sound proof.

In Jiang et al. (2014) an analysis of classification on five layers of available mechanisms for Denial of Service defence has been projected. The mentioned defence mechanisms can prevent, detect, respond to and tolerate the Denial of Service attacks (Jiang et al., 2014). This system provides a stronger understanding of the Denial of Service attack drawbacks and allows a security administrator to combat and mitigate the Denial of Service threat. Denial of Service attack causes disruption of victim's resources in wireless detector networks. As a result of employing this method, a typical detector node isn't ready to access the resources and exhausts its energy. The author has presented a summary of the Denial of Service attack strategy, Denial of Service goals and reasons for Denial of Service over the wireless detector.

There ought to be a link to fault diagnosing of sensible grid with multi-machine language approach. Projected framework overcame the recent challenges in detection of unknown vulnerabilities with lower false positive and false alarms that resulted in higher detection accuracy than the standard IDPSs.

In the IPS White paper (2013), the author declares that the most supporting plan of Intrusion Prevention System is associate inline network-based system. Additionally, another variation of Intrusion hindrance system, layer seven switches has the capabilities for detection and migration of Distributed Denial of Service attack and Denial of Service attack supported awareness of the traffic. Each Intrusion Prevention System can generate associated alert supported policy or signature and can conjointly initiate a response that has been programmed into the system. This alert can happen because of a signature match or violation of individual identification (IPS White paper, 2013).

In Stiawan et al. (2011) the authors have said that in the last few years there has been an explosive growth in the number of attacks on internet with the evolution of new services. Hence, defence system and network monitoring have become essential components of security to predict and prevent attacks. Intrusion prevention system has an additional feature that protects the computer networks. The author has presented mapping problems and challenges related to intrusion prevention system. Further, they summarize the primary and current methods, promising and exciting future directions and challenges in the field related to intrusion prevention system. However, we need to focus on accuracy and precision which is based on behaviour-based prevention (Stiawan et al., 2011).

In Gheorghe et al. (2010), the authors have presented a new security protocol which provides integrity, anti-replay protection, and conversion authentication. To meet these requirements, authors have used MAC and an authentication handshake protocol. The protocol is implemented on Tiny OS in two layers of communication. A protocol has been tested in various scenarios and has proved that the system can reject malicious attempts to communicate with network nodes. The method does not make a difference between an altered packet and a fake MAC and also provides intrusion prevention (Gheorghe et al., 2010).

In Yi et al. (2012) the authors have developed an energy efficient intrusion prevention mechanism in wireless sensor networks which is called as green firewall (Yi et al., 2012). The developed system can isolate an intruder with less overhead and track the intruders continuously to prevent further attacks. The authors have also performed analysis of the overhead cost of the green firewall and compared it with flooding broadcast method. By report, a green firewall can prevent the attack and effectively reduce redundant alarm packet transmissions which result in less energy consumption. The results show that the green firewall can provide low control overhead by reducing the number of alarm packet transmissions; a green firewall can effectively reduce the energy consumption and make it an energy-efficient intrusion prevention mechanism in wireless sensor networks (Yi et al., 2012).

In Rajkumar & Vayanaperumal (2014) the authors have proposed a leader-based intrusion detection system to detect and prevent Denial of Service attacks, Sybil attack and sinkhole attack, by deploying leader-based intrusion detection systems into the access point in the network. The proposed approach can engage security barricades like authentication, favourable incentive provision and Denial of service attack prevention. Further, the same system can do IP verification and packet verification. The simulation is carried out using the Network Simulator 2 (NS2). Authentication, trust calculation, and formal verification are the three primary functions applied to decide whether a node is an intruder node or not. If it is a node, it never allows the nodes for communicating with the other nodes as inter or intra network. From the simulation results, the proposed approach proved that its efficiency is better, and it fulfils the quality of service in the network (Rajkumar & Vayanaperumal, 2014). Literature surveys related to intrusion prevention are as follows:

In Chanaky (2017), the authors have presented the theoretical & practical model for an IPS based on machine learning for detection & prevention of known as well as zero-day attacks. By using the concept of one-class classifiers, they presented an anomaly detector which has detection accuracy for port scan attacks. If there is an intrusion to the system, the system raises an alarm (Chanaky, et al., 2017). Finally, the method is entirely automated, and by using the same system, an organization can save money which is paid to the network administrator.

The proposed system focuses on port scan attack, so there is a scope to develop a system that will work for all kinds of attacks. By using anomaly-based detection, once the unknown attack is detected there is need to update the rule base, but the time required to update the rule base is more than the rate of incoming packets, which can cause the system to slow down and increase its latency. By using the anomaly-based approach, it is possible to detect only unknown attacks. Again, intrusion prevention is not possible by using the same method.

It is challenging to select a system for implementation of intrusion detection over the other. By using learning-based detection systems, if a credible amount of standard traffic data is not available, the training of the techniques becomes very difficult. Intrusion detection algorithm should consider the complex properties of attack behaviours to improve the detection speed and detection accuracy.

In (Mahurkar & Athawale, 2016), the authors state that threats may be identified only with the availability of full data of the system and might be prevented with the employment of correct mechanism or tools, and threats may be avoided with the assistance of appropriate tools or techniques. In addition to wireless LAN the paper also conjointly reviews existing solutions to the threats that exploit communications through wireless fidelity (Mahurkar & Athawale, 2016). Also new prevention mechanisms are presented to address the identified drawbacks by securing its primary server. The paper reviews the experiments conducted to evaluate the impact of varied attacks on the system and conjointly post-resolution results analysis (Mahurkar & Athawale, 2016). A hybrid technique is employed for deep packet scrutiny over network traffic to resolve issues associated with unauthorized access to wireless fidelity resulting in satisfactory performance (Mahurkar & Athawale, 2016). The agent-based communication over wireless network together with the centralized server has improved performance, backup facility and security, by implementing effective security policies and Redundant Array of Inexpensive Disks (RAID) system.

However, the proposed intrusion hindrance system is not able to scale up in case of network overload. The centralized server makes the system stronger in terms of security on a wireless network. However, the centralized server by itself is incapable of addressing each attack.

In Yousufi, Lalwani & Potdar (2017), the authors have alleged that the security of a system or network is one of the most essential parts to keep personal data secure. Most of the attackers try to access the system or network and after that take control over any internal system by exploiting one of them thereby making it is easy to control any other system in the network. A Network Based Intrusion Detection Prevention System (NIDPS) system can inspect both incoming and outgoing traffic in depth. Hence, it prevents attacks that target hundreds of systems inside a network. The authors have designed and implemented a multimode operation IDPS system that takes multiple counter actions against network attacks. At first, it logs the malicious packets, but if the number of malicious packets increases per second, it blocks the attacker's IP address and finally, if it is still unable to prevent the attacker then the corresponding service is stopped so as to make sure the attack does not succeed. By this, the authors have successfully prevented Denial of Service and Brute force attacks launched against File Transfer Protocol (FTP) and Web servers.

However, the system is not able to detect and prevent any type of network attacks including a zero-day attack. If the sleep time of a program is increased, the number of logs crosses 100 eventually causing the service to stop.

In (see GB Hackers in reference), the authors have analyzed some present-day attacks in the network, affecting all layers of the Open Systems Interconnection (OSI) communication model enabling us to understand the security fallacies. The authors have also mentioned mitigation techniques to prevent such attacks. The authors have also introduced various issues related to wired and wireless media that fall into the same categories of attacks along with the offences associated with OSI model layer-wise from top to bottom.

As an extension to the proposed idea it is required to propose an intrusion prevention system for wired and wireless media that will work for cross-layer platforms. In the existing mechanism, attacks to wired media and broadcast media that belong to the same category can only be solved; however, it is unable to prevent an attack that belongs to a different group.

In Shoaib et al. (2017) the authors have constructed a framework known as GDPI. The planned framework works for signatures primarily based on deep packet scrutiny mistreatment of Graphic Processing Units (GPUs). Later the framework is employed to scan the incoming packets to find malicious information within the payload packet by exploitation of some created signatures. Any pattern matching algorithmic program can be used with the developed framework to monitor and implement performance enhancements. The authors have made the framework open for discussion and work. The analysis community is regularly engaged on novel strategies and practices for deep packet scrutiny.

To boost the performance of a system concerning accuracy, there is a requirement to style the system by using the idea of deep learning for intrusion detection and interference. The framework is tested on heterogeneous urologist architectures of NVIDIA GPUs like Tegra Tk1, GTX 780 and Tesla K40. The experimental results achieved a greater speed in packet processing once tested over Tesla K40 with Rabin Karp algorithmic program.

In Sachan (2013), the authors have proposed a novel cluster-based intrusion and detection techniques for misdirection attack. Here, the network parameters are calculated by using the proposed technique resulting in a considerable amount of improvement in throughput while introducing a small amount of delay. The presence of misdirection attack affects the entire performance of the network especially in terms of throughput and end to end delay. The proposed cluster-based intrusion detection and prevention technique is advantageous to detect and prevent misdirection attack. However, the system works only with a given topology.

In Mouradian et al. (2014), the authors have proposed a mechanism which is used to reduce the number of redundant alarms. Here, the nodes compete with each other to be chosen as a sender of alarms. Further, the proposed mechanism has been evaluated by extensive simulations. The author justifies that the scheme can reduce the number of redundant signals, while still refrains from interference with the MAC and routing processes thereby guarantying timeliness and reliability to anomaly detection applications.

In Gandhimathi & Murugaboopathi (2016) the authors have proposed a detection technique that can maximize the detection accuracy by using cross-layer features and mobile-based approach in two phases. In the first phase, attacks are detected by correlating the cross-layer features like MAC and network layer. In the second phase, if the attacks are identified, the mobile-based technique is used to prevent the detected attack. Here, a mobile agent is used to forward the data. Further, the mobile agent is applied to solve problems related to security. By using a mobile-based approach, the false positive rate is reduced, and energy efficiency is improved. The proposed technique efficiently detects and prevents the attack like sinkhole, wormhole and Sybil attack.

In Kalnoor & Agarkhed (2016), the authors have declared that the wireless nature of the network making it conveniently deployable in anywhere within the setting makes it lot more vulnerable to attacks thereby destroying the system. Most significantly, WSN ought to be protected against malicious activities from the pace of deployment and elsewhere, making security a topmost addressable issue. However, such protection is provided by applying mechanisms such as intrusion detection and hindrance. In the proposed system, pattern matching is one amongst the tools, which is used to implement security in wireless sensor network. They conjointly discuss a number of the preventive measures, which will be thought of for safeguarding the detector network. The system tries to search for patterns employed by the various users and matches those patterns with the models that area unit holds within it as information. Upon identification of exact matches, an alert is generated as per the safety considerations. This method effectively safeguards the system from intruders.

In Athawale & Pund (2017), the authors have explained the practical intrusion prevention system, which can measure threats and then takes a decision. Inline intrusion prevention system and passive intrusion detection system presented a roadmap of next-generation intrusion prevention system which can satisfy the needs of future computing. Finally, the authors have presented experimental results for the real-time generated data, which confirms the investigation of IP and MAC. However, this hybrid wireless intrusion system needs innumerable computations, and it results in overhead due to the necessity of periodic searches. Nevertheless, it is agreed that there is a trade-off between the security required and its cost. However, signal frame spoofing can lead to synchronization and is vulnerable to power saving mode attack. These frames cannot be genuine. Hence there is a need to develop an economical approach, which can discover the signal frame spoofing additionally over the entire network.

Table 8.3 shows the gap analysis related to IDS & IPS. From the literature survey it is evident that the existing technologies do not safeguard the system against many attacks like wormhole, Sybil, black hole etc. Hence our goal is to design a system that will address the above mentioned attacks as well.

Table 8.3 Gap analysis IDS and IPS

Sr. No	Reference Paper	Approach Used	Attack Detection	Detected Attack	Prevention
1	Moon, Kim, & Cho, 2014	An energy efficient routing method	Yes	Wormhole attack, Sybil attack	No
2	Athanasios, 2011	SenSys: A lightweight intrusion detection framework	Yes	Sinkhole attack, Wormhole attack	No
3	Chanaky, Kunal, Sumedh, Priyanka, & Mahalle, 2017	SVM: Anomaly-based detection	Yes	Zero-day attack	No
4	Yousufi, Lalwani, & Potdar, 2017	A Network-based intrusion detection system	Yes	Denial of Service, Brute force attacks	No
5	Shoaib et al., 2017	Signature based deep packet inspection using GPUs	Yes	Denial of Service	No

(Continued)

Table 8.3 *(Continued)*

Sr. No	Reference Paper	Approach Used	Attack Detection	Detected Attack	Prevention
6	Gheorghe et al., 2010	Authentication and Anti-replay Security Protocol (AASP)	Yes	Anti-reply attack	No
7	Sachan (2013)	Cluster-based intrusion detection and prevention technique	Yes	Misdirection	No
8	Rajkumar & Vayanaperumal (2014)	A Leader Based Intrusion Detection System	Yes	Denial of Service	No
9	Gandhimathi & Murugaboopathi, (2016)	Novel agent-based approach	Yes	Sinkhole attack, Wormhole attack, Sybil attack	No

8.6 Attack Modelling

Attack modelling is more like procedural flowcharts except that all the activities are uniquely associated with objects. Attack modelling supports the description of parallel operations. Knowledge about how attacks are carried out can help us to develop mechanism that prevents them. Attack modelling is used to describe how activities are coordinated. As per the literature survey, it is found that there is no single paper available that depicts the modelling.

8.6.1 Attacks at Network Layer

The network layer is very essential to send any data. It consists of different routers and many other supporting devices. The primary function of network layer is to deliver each packet from its source to destination (Rajwal et al., 2013). When a packet is in transit the packet is vulnerable to many types of attacks. The various types of attacks that are possible are as follows (Pawar & Anuradha, 2015):

- **Sybil attack:** Sybil attack is a kind of security threat, when a single node in a network claims multiple identities. Many networks rely on some assumptions of character, where each computer represents single identity. A Sybil attack happens when an insecure computer is hijacked to claim multiple identities. Problems arise when a reputation system for ex. file-sharing reputation is tricked into thinking that an attacking computer has a disproportionally large influence (Howard, 2019).

  Figure 8.3 shows a Sybil attack. An assailant initiates a Sybil attack on the malicious node. Then a malicious node establishes unique channels to all its 'n' neighbours broadcast some message. The malicious node is employed to get identities with a random price or steal one from amongst the legitimate nodes. If the channels square measure is valid, the nodes ranging from one to three listen and update their routing tables. Some identities square measure may be dead or out of reach. However, the neighbour nodes of the malicious node notice that each of their encircled nodes square measures are alive and approachable (Deshpande et al., 2018).

- **Wormhole attack:** In the wormhole attack, an attacker receives the data packet at one location in the network and tunnels them to another site in the network. The tunnel present between two devices of an attacker is called a wormhole. Wormhole attacks are a severe threat to ad hoc network routing protocols. When the wormhole attacks are used by the

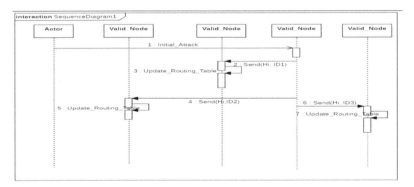

Figure 8.3   Sybil attack.

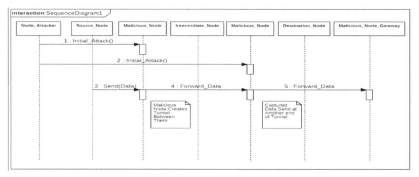

Figure 8.4   Wormhole attack.

Source: (Howard, 2019)

attacker in the routing protocol, the attack prevents the discovery of any path other than through the wormhole (Deshpande et al., 2018).

Figure 8.4 shows how the wormhole attack takes place. An assaulter initiates wormhole attacks on the malicious nodes. A malicious device sends salutation messages to its neighbours to update their routing table thereby determining the destination to the malicious device. The neighbour device of the malicious device transmits information whenever an event occurs. The malicious device receives information from the neighbours and forwards it to other devices by establishing a tunnel between the source device and the victim (Deshpande et al., 2018).

- **Blackhole attack:** In these attacks, the attacker claims to have an optimal and authenticate route to a destination device for forwarding the packet to a destination. Attacker accepts the request and sends the fake reply through the short way; thus, phony route is created. Once the attack takes place in between the communicating devices, the attacker can use the packets to his convenience and achieve his goal.

Figure 8.5 shows the detailed steps in the Blackhole Attack. An attacker compromises a malicious node and initiates a Black hole Attack on the malicious node. Source node detects an event and then forwards the event data to the intermediate node that is on the routing path. The intermediate node transmits the received data from the source node to malicious node by following the routing path. A malicious node may refuse to forward specific messages and just drop them, ensuring the termination of propagation.

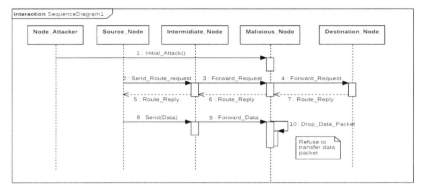

Figure 8.5   Blackhole attack.

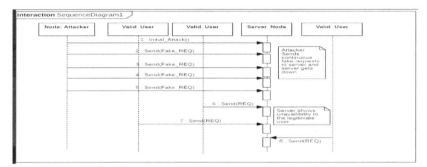

Figure 8.6   Denial of Service attack.

- **Denial of Service attack:** Denial of Service attack is used to prevent access to the authorized user. By sending multiple requests to the host, DoS is implemented. If the server is not able to handle multiple requests, then the server becomes overloaded and becomes unavailable to to service the requests from a legal user.

  Figure 8.6 shows the denial of service attack. In this, an attacker uses valid nodes and then starts the denial of service attack. An attacker sends fake requests continuously to the server node until the server node becomes unavailable to service the requests.

8.7 Issues and Challenges

Now-a-days, security is one among the foremost important aspects of any system, and folks have a special perspective concerning security. Generally, security may be thought of as the safety of a system as a whole.

The communications in wireless detector network applications are largely wire-less in nature. This could end in varied security threats to such systems. These threats and attacks may cause severe threats concerning the life of a personal UN agency by victimization of the wireless detector devices like smart-phones. In some cases, following the placement of an individual if compromised could cause grave consequences. Folks with malicious intent could use the personal knowledge to hurt the person. Security problems in applications of wireless detector networks have continuously been a part of the active analysis. Security problems, generally in wireless detector networks provide a wide avenue for analysis in recent times.

8.7.1 Why are Sensor Networks Difficult to Protect?

Wireless sensor network is very difficult to protect because of limited resources, unreliable communication, Ad-Hoc Deployment, Immense Scale, Unattended Operation, and security Requirements.

8.7.2 Issues in Wireless Sensor Network

Security design for networks should integrate some security measures and techniques to guard the network and satisfy the required needs. To realize a secure system, security should be incorporated into each part/components of the network. Security should perforate to each side of the underlying system style.

As wireless device networks are unceasingly growing, the requirement for adequate security mechanisms are also growing with them. Since device networks typically move with sensitive knowledge and operate in hostile unattended environments, it's imperative that these security considerations be self-addressed from the start of the system style. To combine all the mandatory parts (sensors, radios, and CPUs) into an energetic device network needs a short understanding of capabilities and limitations that come as a package. Every node should be designed to supply the set of primitives necessary to synthesize the interconnected topology because it is deployed to meet strict needs of size, cost, and power consumption.

8.7.3 Challenges in Wireless Sensor Network

Now a day, security is one of the most critical aspects of any system, and people have a different perspective regarding security, and hence it is defined in many ways. In general words, security is a concept like the safety of the system as a whole.

The communications in wireless sensor networks applications are mostly wireless in nature. This may result in various security threats to these systems. These threats and attacks could pose severe problems to the social life of an individual who is using the wireless sensor devices like smartphones. In some cases, tracking the location of a person if compromised may lead to grave consequences. People with malicious intent may use the private data to harm the person. Security issues in applications of wireless sensor networks have always been part of the active research. Security issues in general wireless sensor networks are a significant area of research in recent times.

8.8 Proposed Work

Typically, a firewall is used to inspect the header and compares it with the firewall's policy which then decides on the action to be taken in terms of acceptance or rejection. The intrusion detection system is used to examine the header only. But our proposed system monitors the header as well as the payload using deep packet inspection mechanism with which required action will be taken to prevent the attack. So, an Intrusion Prevention System is crucial to prevent a threat from penetrating networks.

Figure 8.7 shows the IPS architecture to prevent attacks in a wireless sensor network. The system depicted in the architecture starts its functioning by capturing packets as an input, followed by data analysis based on which a response is given. Finally, a filtered packet is provided as an output.

So, there is a need to design a system to prevent intrusion in a wireless sensor network. Due to anything anytime anywhere type of computing, use of wireless sensor network has increased drastically. Looking at the various threats and several attacks in a wireless network, security is of prime concern. Intrusion is one of the critical issues, and intrusion detection has already reached its saturation. Therefore, we need an efficient solution for proactive intrusion prevention towards wireless sensor networks.

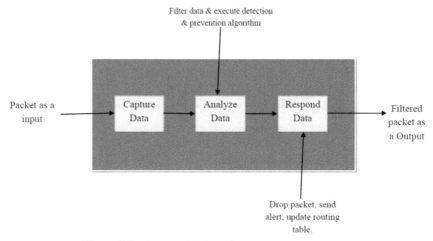

Figure 8.7   Proposed IPS architecture to prevent attacks.

8.9 A Need of Today

The next section deals with the need of today and impact of adopting or not adopting the proposed approach.

8.9.1 What should be the Adverse Implications for not Adopting this Approach

If we are not going to adopt the provided approach, then definitely there should be some loss in terms of data, personal information, valuable assets and many more. Let us think about a straightforward example in which we are not going to adopt proposed approach, within which we tend to a home and two-three peoples are staying reception. Once more assume all the peoples staying therein home are operating in their workplace in between 11.00 am to 5.00 pm. Therefore, no one is available at home in same timing and surely that home is locked. Definitely, unknown person will enter the house in between 11.00 am to 5.00 pm and might do banned things similar to theft. As a result of that theft, some valuable assets from same home will be lost.

Quoting the example presented in Section 8.1.4 it is evident that a property/household without security measures is prone to theft.

Adoption of the proposed approach prevents the loss of valuable assets which can be implemented by appointing a security guard. The security guard will then restrict the entry of unknown people.

8.9.2 What should be done by Industry and Policy Makers?

Industries can provide some guidelines in terms of standardization. Standardization facilitate sharing of knowledge and best practices by helping to ensure common understanding of concepts, terms, and definitions, which prevents errors. Standardization helps to establish common security requirements and the capabilities needed for secure solutions. For example, Federal Information Processing Standards (FIPS) 140-2, Security Requirements for Cryptographic Modules, establishes standard requirements for all cryptographic – based security systems used by federal organizations to protect sensitive or valuable data. Other helpful resources include the Internet Security Glossary from the Internet Engineering Task Force (IETF), Request for Comments (RFC) and a compendium of the International Telecommunications Union Telecommunication Standardization Sector (ITU-T)

Policy makers should design some policies by involving stakeholders. Policy makers can back up the design by conducting some awareness drives.

8.10 Conclusions and Future Work

From the discussions in the above sections we understand the need to design a system to prevent intrusions in a wireless sensor network. Due to anything anytime anywhere type of computing, use of wireless sensor network has increased drastically. Looking at the various threats and several attacks in a wireless network, security is the prime concern. Intrusion is one of the critical issues, and intrusion detection has already reached its saturation. Therefore, we need an efficient solution for proactive intrusion prevention towards wireless sensor networks.

This chapter presents a comprehensive survey and discussion as well as gap analysis on intrusion detection and prevention system for wireless sensor network using machine learning. Also, gap analysis has been presented with regard to attacks at network layer for the wireless sensor network. Through the literature survey we understand that there is a need to develop scalable and attack resistance system for intrusion prevention using deep packet inspection in a wireless sensor network. A system is proposed to detect and prevent intrusion using machine learning for the wireless sensor network.

Hence, the proposed system can be used to detect and prevent attacks in all types of networks where security matters.

References

Ashoor, A. S. and Gore, S. (2001). Difference between Intrusion Detection System (IDS) & Intrusion Prevention System (IPS). *International Conference on Network Security and Applications CNSA 2011: Advances in Network Security and Applications*, pp, 497–501.

Athanasios, G. (2011). *Security Threats in Wireless Sensor Networks: Implementation of Attacks & Defense Mechanisms*. Aalborg University.

Athawale, S. V. and Pund, M. A. (2017). NGIPS: The Road Map of Next Generation Intrusion Prevention System for Wireless LAN. 978-1-5090-4264-7/17/2017.

Chaudhari, H. C. and Kadam, L. U. (2011). Wireless Sensor Networks: Security, Attacks, and Challenges. *International Journal of Networking,* 1(1), pp. 4–16.

Chopra, B. and Singh, P. (2014). A Survey on Different aspects of Network Security in Wired and Wireless Networks. *International Journal of Latest Trends in Engineering and Technology (IJLTET)*, 4(2).

Chanakya, G., Kunal, P., Sumedh, S., Priyanka, W., and Mahalle, P. N. (2017). Network Intrusion Prevention system using Machine Learning techniques. *International Journal of Innovative Research in Computer and Communication Engineering*.

Das, S. and Nehe, M. J. (2017). A Survey on types of machine learning techniques in intrusion prevention systems at *IEEE WiSPNET*, 978-1-5090-4442-9/17/2017.

Deshpande, A. V., Mahalle, P. N., and Kimbahune, V. V. (2018). A Novel Key Management Scheme for Next Generation Internet – An Attack Resistant and Scalable Approach. *International Journal of Information System Modeling and Design archive*, 9(1), 92–121.

Gandhimathi, L. and Murugaboopathi, G. (2016). Cross-layer Intrusion Detection and Prevention of Multiple attacks in Wireless Sensor Network using Mobile Agent. *International Conference On Information Communication And Embedded System* (ICICES 2016).

Gheorghe, L., Rughiniş, R., Deaconescu, R., and Ţăpuş, N. (2010). Authentication and Anti-replay Security Protocol for Wireless Sensor Networks. *Fifth International Conference on Systems and Networks Communications*.

Hackers, G. B. Intrusion-detection-system. Available://www.gbhackers.com.

Hellbusch, S. A. (2014). *Wireless Sensor Network Security*. Carnegie Mellon University, Pennsylvania. Available at: https://docplayer.net/9790 089-Wireless-sensor-network-security-seth-a-hellbusch-cmpe-257.html

Host *vs.* Network-Based Intrusion Detection System, SANS Institute 2000–2005. Available://www.sansinstitute.com.

Howard, A. (2019). What is a Sybil Attack? TopTenReviews. Available https://www.toptenreviews.com/software/articles/what-is-a-sybil-attack/.

Intrusion Detection Systems Buyers Guide. Available://www.intrusiondetectionsystem.com

Intrusion Prevention Systems (IPS). (2003). Part one: Deciphering the inline Intrusion Prevention hype, and working toward a real-world, proactive security solution, White Paper.

Jadhav, R. and Vatsala, (2017). Security Issues, and Solutions in Wireless Sensor Networks. *International Journal of Computer Applications* (0975–8887).

Jiang, Y., Huang, J., and Jin, W. (2014). Intrusion Tolerance System Against Denial of Service Attacks In Wireless Sensor Network. *International Conference on Cyberspace Technology (CCT 2014).*

Kalnoor, G. and Agarkhed, J. (2016). Pattern Matching Intrusion Detection Technique for Wireless Sensor Networks. *International Conference on Advances in Electrical, Electronics, Information, Communication, and Bio-Informatics.*

Kim, K. and Aminanto, M. E. (2017). Deep Learning in Intrusion Detection Perspective: Overview and Further Challenges. *2017 International Workshop on Big Data and Information Security (IWBIS)* 978-1-5386-2038-0/17/2017.

Mahurkar, K. K. and Athawale, S. V. (2016). A New Improve Intrusion Prevention System Security for Wireless LAN A Review. *International Journal of Engineering And Computer Science,* ISSN: 2319-7242, 5(11), 19116–19118.

Moon, S. Y., Kim, J. W., and Cho, T. H. (2014). An Energy-Efficient Routing method with Intrusion Detection & Prevention for Wireless Sensor Networks. *ICACT 2014*, ISBN 978-89-968650-3-2.

Mouradian, A., Nguyen, X. L., and Augé-Blum.Preventing, I. (2014). Alarm Storms in WSNs Anomaly Detection Applications. IEEE.

Pawar, M. V. and Anuradha, J. (2015). Network Security and Types of Attacks in Network. *International Conference on Intelligent Computing, Communication & Convergence, Conference*. Organized by Interscience Institute of Management and Technology, Bhubaneswar, Odisha, India.

Patel, A., Alhussianc, H., Pedersen, J. M., Bounabat, B., Júnior, J. C., and Katsikas, S. (2017). A nifty collaborative intrusion detection and prevention architecture for Smart Grid ecosystems, *Computers & security* 64, 92–109.

Rajwal, D., Band, D., and Yadav, A. (2013). Study of Different Attacks on Network & Transport Layer. *International Journal of Engineering and Computer Science,* ISSN: 2319-7242, 2(3), 692–695.

Rajkumar, D, and Vayanaperumal, R. (2015). A Leader Based Intrusion Detection System for Preventing Intruder in Heterogeneous Wireless Sensor Network. *IEEE.*

Sachan, R. S., Wazid, M., Singh, D. P., and Goudar, R. H. (2013). A Cluster-Based Intrusion Detection & Prevention Technique for Misdirection Attack inside WSN.International Conference on Communication and Signal Processing.

Shoaib, N., Shamsi, J., Mustafa, T., Zaman, A., ul Hasan, J., and Gohar, M. (2017). DPI: Signature-based Deep Packet Inspection using GPUs. (IJACSA) *International Journal of Advanced Computer Science and Applications*, 8(11), 210–216.

Singh, J. and Nene, M. J. (2013). A survey on machine learning techniques for intrusion detection systems. *International Journal of Advanced Research in Computer and Communication Engineering.*

Stiawan, D., Abdullah, A. H., and Idris, M. Y. (2011). Characterizing Network Intrusion Prevention System. *International Journal of Computer Applications* (0975–8887).

Tsai, C., Hsu, Y., Lin, C., and Lin, W. (2009). Intrusion detection by machine learning: A review. *Elsevier Expert Systems with Applications*, 36, 11994–12000.

Yi, P., Zhu, T., Wu, Y., and Li, J. (2012). Green Firewall: an Energy-Efficient Intrusion Prevention Mechanism in Wireless Sensor Network. IEEE.

Yousufi, R. M., Lalwani, P., and Potdar, M. B. (2017). A Network-Based Intrusion Detection and Prevention System with Multi-Mode Counteractions. *2017 International Conference on Innovations in Information, Embedded and Communication Systems (ICIIECS).*

9

Comprehensive Threat Analysis and Activity Modelling of Physical Layer Attacks in Internet of Things

Mahendra B. Salunke[1], Parikshit N. Mahalle[1] and Prashant S. Dhotre[2]

[1]SKNCOE Research Center, SPPU, Pune, India
[2]Department of Computer Engineering, DYPIT, SPPU, Pune, India
msalunke@gmail.com, aalborg.pnm@gmail.com,
prashantsdhotre@gmail.com

Day by day, the use of smart devices by Internet users has been increasing at a rapid pace and has been applied in several places like homes, offices and other environments. The environment with the help of smart devices has not only offered many services to Internet users but also leads Internet users to enter in invulnerable, the untested environment of "Internet of Things" (IoT). With the rapid expansion of IoT, the safety of our embedded devices is at highest risk, which motivated us to do a review of several types of physical layer attacks. Traditionally, the researchers focused on providing security solutions for higher layers like network layer, application layer, etc. Keeping the physical layer at the center point, the paper begins with an overview of IoT, use cases & challenges are presented. However, this paper has presented a discussion on the threats and attacks at the physical layer along with its solutions proposed by the researchers. The main contribution of this paper is a detailed discussion of threat analysis and activity modelling of physical layer attacks. This paper presents several research challenges in embedded security. The paper concludes with the proposed methodology to address research challenges that make the devices much safer and less vulnerable to the attacks.

9.1 Introduction

The term 'Internet of Things' was first coined by Kevin Ashtonin 1999, which is a technological revolution that represents the future of computing and communications (Rose et al., 2015). Looking towards the present expansion of IoT, the demand for connecting devices is found across multiple industries. This section focuses on the current market scenario of IoT, various applications and challenges in it.

9.1.1 Overview

According to NASSCOM, the global market size of IoT will reach up to USD 3 trillion by 2020 (Sudarshan et al., 2017). IoT is supposed to change the entire way of communication, their work and lifestyle. This leads to omnidirectional communication where everyone can connect to everything, everywhere. The entire world is migrating towards IoT, which will have a huge impact on our lives in the coming five years. This will have an influential impact on how the businesses and government interact with the world. IoT is service-oriented architecture and a mandatory subset of the Future Internet, where every virtual or physical thing can communicate with every other thing giving seamless service to all stakeholders (Mahalle et al., 2013a; Mahalle et al., 2013b). Not only computers or laptops or smartphones connect with the internet, but also there will be a huge number of smart devices connected to the internet and in turn to each other. Starting from home appliances to big industrial machinery, everything is becoming intelligent using smart devices and technologies. This is leading to the new smart computing environment called 'Next Digital Revolution' or 'Next Generation of Internet' (Taylor & Boniface, 2013).

IoT consists of billions of tiny devices, people, services, several digital devices and other physical objects to form a collaborative computing environment to extend communication and networking services to anyone at anytime and anywhere. These objects generate a huge amount of data, which are then processed using smart technologies to offer useful services to Internet users in areas like home and building automation, intelligent transportation and connected vehicles, industrial automation, smart healthcare, smart cities, and others (Kanuparthi, 2013). The IoT facilitates different communication patterns like user-to-user, user-to-device, device-to-device, and devices-to-user (Mahalle et al., 2014). The components of IoT are capable to flawlessly connect, interact and exchange information among them to make Internet users' lives simpler and easy through a digital environment. Hence, the environment is becoming sensitive, adaptive and responsive to human needs. To create an IoT, computing platforms are expected to be embedded within the physical components and people. These embedded computing platforms enable a wide range of applications like implantable and wearable medical devices, smart homes/buildings, smart grids, brain-machine interface, intelligent automobile & transportation system, physical infrastructure monitoring, smart meters and many more (Babar et al., 2011). Figure 9.1 shows various components of IoT.

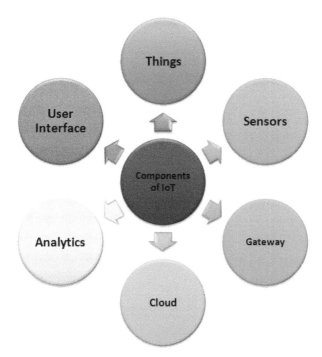

Figure 9.1 Components of IoT.

The fundamental components of IoT are:

- **Things:** Things in the IoT are the physical objects with an embedded system having unique identification and capable to collect and transfer data.
- **Sensors:** Sensors are fixed on the things to collect various parameters associated with the thing or the environment nearby it.
- **Gateway:** This is an interface between things and cloud, which uses one of the existing communication channels (e.g. GSM, Wi-Fi etc.) to establish communication and exchange the information.
- **Cloud:** Cloud allows us to store and process data that is received from various sensors attached to the things.
- **Analytics:** The received data is processed and analyzed to gain valuable knowledge and/or to take necessary actions by means of sending necessary signals to actuators.
- **User interface:** It is a visual interface between the user and the system, which is required to represent the analyzed data and to enable the user to control the system.

9.1.2 IoT Based Business Use Cases

IoT can be defined as a future where all physical objects are connected to the internet, providing various solutions to automate various processes in industries and other segments. IoT can be divided into three categories, based on usage and client-base as Consumer IoT, Commercial IoT and Industrial IoT.

Consumer IoT includes connected devices such as smart cars, phones, watches, laptops, connected appliances, and entertainment systems. Commercial IoT includes things like inventory controls, device trackers, and connected medical devices. Industrial IoT covers such things as connected electric meters, wastewater systems, flow gauges, pipeline monitors, manufacturing robots, and other types of connected industrial devices and systems.

There are several sectors where IoT technologies play a vital role. Figure 9.2 gives a brief idea about the business scenario using IoT (El Mouaatamid et al., 2016).

- **Smart Home:** The main objectives of smart homes are to provide security, comfort and perform resources management. Smart homes are extended a comfort zone to the Internet users where a homeowner can get detailed information about the availability of vegetables in the fridge

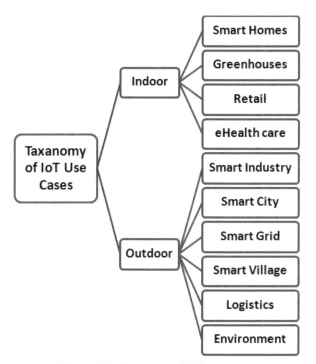

Figure 9.2 Taxonomy of IoT use cases.

so that a user can order it online from office. There are several other solutions like adjusting light system according to the user's mood, brewing coffee based on the user's arrival etc. makes use of heterogeneous devices to implement a smart home.

- **Greenhouses:** Due to the demand for best quality agricultural productions, greenhouses are built to provide a controlled environment for better quality and quantity. Using IoT solutions, the farmers can monitor the internal dynamic parameters, can automate the necessary actions like watering, giving fertilizers, spraying pesticide etc., and can plan the harvesting schedule.
- **Retail:** IoT plays a very important role in the retail segment to improve business and to meet customer's demands.
- **Smart Industry:** Industrial automation is the major demand of managing committee to minimize the production cost and to improve the quality and quantity of products, to manage the infrastructure and human resources etc.

- **Smart City:** In the urban region, local governing bodies can deploy various solutions using ICT to manage the services and resources where they can manage infrastructure (like water, energy, transportation, waste management etc.) and citizens can avail best possible online services to pay various bills, to raise the complaints, to provide feedback of services availed etc.
- **Smart Grid:** It is associated to supply of various resources like water, oil, electricity, gas etc. and making these resources available on demand. Smart grids help to manage the bills and providing an option to pay bills online etc.
- **Smart Village:** Due to expansion in telecommunication infrastructure, the ICT is available in almost all villages where the latest generation networks are available. With the advancement in the development, this segment is also making use of IoT technologies to address various local issues like irrigation, identification of disease on crop, planning harvesting based on market demand, filing online application forms to get benefits of government schemes and tracking the status of the same etc.
- **Logistics:** Intelligent Transportation System (ITS) applications is one of the examples of ICT in logistic. Due to the features to manage the complete supply chain, these systems are intelligent to monitor and track the things online.
- **Environment:** Using various sensors and actuators the system monitors the environment and manage the environment as required. These types of solutions are useful to provide emergency alerts and take precautionary measures.

9.1.3 Challenges in IoT

Even though IoT offers valuable services in the various sections, it is coming with a few important issues and challenges. There are around six challenges in the IoT as mentioned in a research work (El Mouaatamid et al., 2016). The challenges are explained with the example of e-Health solution.

- **Sensing a complex environment:** The challenge is to provide innovative ways to sense the parameters in a given environment and deliver to the cloud via the physical world. In e-health application, heterogeneous sensors are fitted on the human body as well as implanted inside the body to sense different body parameters.

- **Connectivity:** To establish a flawless connection between communication entities by using various available wired and wireless standards is another challenge. In e-health application, the parameters sensed are to be transmitted to controllers for further processing, which may use a wired or wireless communication system.
- **Power:** The other challenge is to ensure availability of power of the devices throughout the life span in the IoT environment. In e-health applications, the implanted sensors need enough power so that they can function properly throughout its life span.
- **Security:** To detect and protect users' data from attacks are the key features required to ensure the security of the system. In addition, the privacy protection is another challenge that IoT solutions must consider. In e-health application, patient data is sensitive and need to be protected from an illegitimate user.
- **The complexity of application development:** The application development must be easy so that those can be designed and developed by all types of developers (be they inexperienced or experienced). In the e-health application, application development may be complex due to the complex medical field. The applications need to be developed so that they can support medical practitioners to understand the patient's conditions and decide the further action plan.
- **Cloud services:** The challenge is to provide reliable cloud services. In the e-health application, the reliability of cloud services is the most important requirement as the application is directly related to the patient's life.

Having a discussion on several challenges in IoT, this paper focuses on research contribution on securing the IoT system.

9.2 Motivation

IoT is a network system of embedded objects or devices, with a unique identifier, in which communication without any human intervention. This could be possible using standard and interoperable communication protocols. Embedded devices are an integral part of IoT and play a vital role in providing computing, storage, connectivity and communication facilities. The explosion in devices and connectivity creates a much larger attack surface, opening new opportunities for malicious people and entities. Unless significant attention is paid to security, the IoT could well be turned into an

Internet of "Things to be Hacked!" (Kermani et al., 2013). As the application areas of safe embedded systems evolve, handling sensitive private data and assuming critical roles, their requirements change accordingly, and certain security issues become more persistent (Charalampos et al., 2014).

To explore attack surfaces in detail, let us take a virtual shopping scenario.

> *Ramesh is working in a multinational company and he is posted to a different continent than his own country's continent. On the 20th birthday (which is due in the next month) of his only son Aniket, he wants to gift a multipurpose car so that Aniket can drive the car to his college as well as he can take family members for shopping or to small trips. To select the vehicle Ramesh called a showroom located in his hometown and expressed his willingness to purchase a vehicle. Sales executive want to give the best option for which he requests Ramesh to give access to an IP based CCTV camera installed in the garage to estimate the parking space available and to see the bungalow colour to finalize the vehicle colour. After estimating available space and bungalow colour, executive gives him few options to select. After confirming the order, Ramesh pays the bill using net banking. The delivery manager checks with the availability of the ordered model and confirm the delivery schedule for a given date and deliver the selected vehicle to Ramesh's home.*

In Figure 9.3, we can easily identify the attack surfaces as a link between devices and IoT services, device low-level software, attacking the things and gateways.

- **Role of IoT and security concerns in India**

Today due to increasing demand for quality of service to improving the standard of life, consumers are demanding best services from government, semi-government and private service providers. To improve infrastructure required to provide the best services, Ministry of Urban Development, Government of India selected 100 cities under Smart Cities Mission (see Ministry of Urban Development Government of India, 2015). The main objective of this mission is to promote cities that provide core infrastructure and best quality of life to its citizens, a clean and sustainable environment and make use of 'Smart' solutions. The core infrastructure elements are adequate water supply, assured electricity supply, sanitation-including solid waste management, efficient urban mobility and public transport, affordable housing for the poor, robust IT connectivity and digitization, good governance

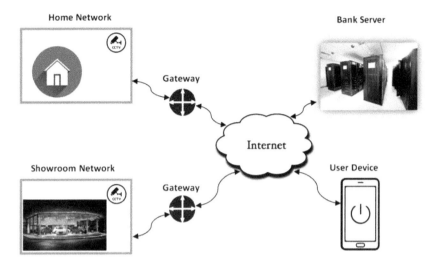

Figure 9.3 Virtual shopping scenario.

(e-governance and citizens participation) sustainable environment, safety and security of citizens (women, children and the elderly) and health and education. To achieve objectives, it becomes necessary to use technology to large extent, especially technology that achieves value addition in the services and gives smart outcomes. To improve infrastructure and services, the application of Smart Solutions will enable cities to use technology, information and data. Due to this need, a huge number of IoT based smart solutions are in demand.

Following are a few examples where smart solutions using IoT are proposed in smart cities mission (Ministry of Urban Development Government of India, 2017):

- Automatic Vehicle Locating system
- Integrated command and control centre
- Smart lighting
- Smart parking
- Public bike sharing system
- The intelligent traffic management system
- Transport command and control centre (real-time tracking of 1500+buses)
- Integrated CCTV surveillance of street
- Jan-Mitra Card for non-transit services
- And many more

Though we are utilizing the benefits from IoT that wireless connection has brought, its broadcast nature makes the transmission vulnerable to passive eavesdropping and active jamming (Zhang et al., 2017). The data transmitted in the IoT may contain sensitive, private or confidential information, hence it becomes necessary to protect it on-the-fly. For example, in home automation applications, the sensor nodes collect real-time data from the home like temperature, humidity and Closed Circuit TV (CCTV) footage (inside and outside the home), etc. This has been observed that the devices are vital This information is private, sensitive and valuable. Hence, a secure transmission channel is required.

Due to the digitization of several operations in industrial as well as domestic applications, IoT is becoming an embedded entity everywhere especially in Indian context (Das, 2017).

- **Skilled workforce:** Due to the sophisticated system, it becomes a great challenge of providing training to existing manpower to acquire necessary skills or hire skilled manpower. The employees should understand the security concerns related to IoT operations. Also, the relevant certifications on the IT and IoT parts of the organization become necessary. In addition to this, it becomes mandatory to provide additional and incremental training to current employees, which may help scale up them in the IoT domain.

- **Security:** Cybersecurity is the main challenge while adopting IoT based solutions, as each object is connected to the internet, which increases potential entry points for attackers. Due to the increased attack surface, hackers can easily hack the system and may cause damage, which may be life-threatening. In India, Internet users are using many devices and carrying them wherever they go. This leaves the footprints for the attackers and creates possibilities of attack causing a loss to the individual users or group users.

   Using multilevel security, we can easily secure the IoT. The security mechanism should be used at three different levels viz. at device and sensors, at the network and at the application level. For better security, all these three levels should have a tamperproof security mechanism. There are security solutions proposed by different researchers for network and application level, but less work is carried out at the device and sensor level. Due to the resource-constrained nature of endpoints, at device and sensor level, it required to design and develop a lightweight embedded security mechanism.

- **Privacy:** Due to the rapid expansion of IoT, every individual is making use of services offered by the system. The system collects user information, monitors and records various parameters around the user, due to which it become necessary to protect the user's rights to the privacy so that the system can be treated as trustworthy. The applications like eHealth care, surveillance system, smart homes, e-governance, transportation management etc. handles users' confidential data and due to openness of networks used in IoT, unauthorized access to this data becomes more challenging. The IoT system adheres to the privacy protection principles that includes consent, limited information collection, notification, etc. (Cavoukian & Reed, 2013).
- **Lack of standardization:** Due to the isolated nature of researchers and manufacturers of IoT, there are no global standards, especially in India. Due to the heterogeneous nature of devices deployed, interoperability becomes another challenge. Hence, it becomes necessary for researchers to come on a common platform, discuss and finalize the standards required for IoT. Standardization of technologies will lead to better interoperability, and thus lowering the costs. Also due to interoperability, it will become easy to add and remove the sensor and device in the IoT system.

9.3 Related Work

Considering the security in the discussion, this section gives extensive state-of-the-art technology. This part deals with the various physical layer attacks and security solutions proposed by different researchers.

In their paper presented by Junqing Zhang et al. (2017) overviewed the latest research efforts on different approaches to secure a widely used communication technique in IoT i.e. wireless communication, at the physical layer. The authors emphasized that the design of a low-cost and robust cryptosystem for IoT is vital and proposed the main security streams have on the upper layers. However, the physical layer can also be leveraged to enhance security, which may reduce additional energy cost for security.

Sachin Babar (2015) addressed the issues in embedded security for IoT and developed the mechanism to protect IoT from a jamming attack. The proposed security framework for IoT is a collection of contributions in the threat taxonomy, attack modelling, attack detection, key management and defence mechanism. This work is more efficient in terms of energy consumption, delay and throughput.

To improve security at the physical layer, Ying Bi & Chen (2016) developed a new cooperative protocol named as Accumulate-and-Jam (AnJ). To secure the direct communication between two terminals in the presence of a passive eavesdropper, a Full-Duplex (FD) friendly jammer is deployed. The friendly jammer is an energy-constrained node without an embedded power supply but with the energy harvesting unit and rechargeable energy storage, which harvest energy from the radio waves transmitted by source and store in its battery to perform its basic operation of cooperative jamming. Using proposed AnJ protocol, the system operates either in Dedicated Energy Harvesting (DEH) or in Opportunistic Energy Harvesting (OEH) mode based on the energy status of the jammer and the channel state of the source-destination link. In DEH mode, the jammer performs energy harvesting using energy-bearing signals sent by the source. In OEH mode, the source transmits an information-bearing signal for the destination node and at the same time, using the harvested energy, jammer transmits a jamming signal to confuse the eavesdropper. Due to the FD capability, the jammer also harvests energy from the information-bearing signal. In this paper, authors investigated and discussed the dynamic charging/discharging behaviour of the finite capacity energy storage at the jammer, and the secrecy outage probability and the existence of the non-zero secrecy capacity of the proposed AnJ protocol. Using a wireless-powered half-duplex jammer, authors also derived the secrecy metrics half-duplex jammer to serves as a performance benchmark. The obtained numerical results demonstrated that the proposed protocol could provide not only a superior performance over the conventional half-duplex schemes but also a satisfactory performance close to the upper bound when the energy storage is sufficiently subdivided.

After analyzing different threats in the IoT, Tommaso Pecorella et al. (2016) proposed a security framework for the device initialization and demonstrated the effectiveness of physical layer security in the IoT systems. In this paper, authors addressed the problem of securing the configuration phase of an IoT system and highlighted on the main drawbacks of existing solutions, in which focus on specific techniques and methods and lack of cross-layer approach. The advantages and disadvantages of the Advanced Encryption Standard (AES) algorithm and Elliptic Curve Cryptography (ECC) are elaborated and finally proposed to use ECC cryptographic system. The authors classified the existing physical layer security approaches based on the physical characteristic exploited viz. secrecy capacity, channel signature/fingerprint, spectrum spreading of signal energy and cooperation.

Ning Wang et al. (2017) addressed the physical layer spoofing attack in wireless communication. In that solution, the authors proposed a physical-layer spoofing detecting scheme to improve the detection performance, where signal processing and feature recognition are utilized. To reinforce the characteristic of the signal, the authors presented a pretreatment process based on sparse representation (SR). They have formulated the problem of spoofing detection as one of the feature extractions and recognition and employed a fuzzy C-mean algorithm to further increase the recognition accuracy.

Considering the need fora lightweight security mechanism for M2M communication in Industrial IoT, Alireza Esfahani et al. (2017) proposed a lightweight authentication mechanism using hash and XOR operations. The observations indicate that the proposed mechanism is effective in terms of computational cost, communication and storage overhead and is resistant against replay, man-in-the-middle, impersonation and modification attacks.

Zhicai Shi et al. (2017) proposed a lightweight authentication protocol based on CRC function which needs less computation and storage compared to Hash functions. As per the observations, this protocol exploits an on-chip CRC function and a pseudorandom number generator to ensure the anonymity and freshness of communications between the RFID reader and tag. This protocol is most effective in preventing eavesdropping, location trace, replay attack, spoofing and DOS- attack on the RFID system.

Yi-Sheng Shiu et al. (2011) highlighted the most common attacks seen in wireless communication and categorized broadly as active and passive attacks. The passive attacks often used are eavesdropping intrusion and traffic analysis. On the other side, the active attacks include Denial-of-Service (DoS) attacks, masquerade and replay attacks, and information disclosure and message modification attacks. They have classified the existing physical layer security methods into five major categories: theoretical secure capacity, channel approaches, coding approaches, power approaches, and signal detection approaches. To improve security in wireless transmissions, the authors have introduced and compared the existing physical layer security approaches in terms of their abilities and illustrated the effectiveness of some physical layer security schemes.

In the context of wireless sensor-sink transmissions in industrial wireless sensor networks (IWSNs), Zhu et al. (2017) evaluated the tradeoff between security and reliability, called "security-reliability tradeoff". In this paper, the authors presented a review of the challenges and solutions for improving the physical-layer security and reliability for IWSNs. For mitigating the background interference, path loss, multipath fading, and link failure, several wireless reliability enhancement techniques are discussed. The paper also includes an overview of wireless jamming and eavesdropping attacks along with their countermeasures.

A survey paper on potential security issues in existing wireless sensor network protocols has been presented by Tomic and McCann (2017). This work has carried out a thorough analysis of the main security mechanisms and their effects on the most popular protocols and standards used in WSN deployments i.e. IEEE 802.15.4, B-MAC, 6LoWPAN, RPL, BCP, CTP, and CoAP. At each layer of the WSN stack, the authors have presented the potential security threats and existing countermeasures. The authors concluded that the broadcast nature of wireless communication makes the system susceptible to jamming, eavesdropping, node tampering, and hardware hacking at the physical layer.

A work 'DTD: A Novel Double-Track Approach to Clone Detection for RFID-enabled Supply Chains' is a novel approach for effective clone detection has been presented by Jun Huang et al. (2017). In this paper, verification information is written into tags as a part of its attribute, which forms a time series sequence during tag verification. Based on the discrepancy in the verification sequences, cloned tags can be easily differentiated from genuine tags. The theoretical analysis and experimental results show that the proposed approach is more effective, reasonable and has a relatively high clone detection rate.

In context to IoT, Amitav Mukherjee (2015) presented an overview of low-complexity physical-layer security schemes. The authors presented a local IoT deployment model, which consist of multiple sensor and data sub-networks having sensors to controllers' uplink communications, and controllers to actuators downlink communications. Also, the paper shows a review of the state of the art in physical-layer security for sensor networks, followed by an overview of communication network security techniques. The work also emphasizes the need for a well-founded and holistic approach for incorporating complexity constraints in physical-layer security designs.

Another work has reviewed several flavours of physical layer security specifically for wireless systems by Wade Trapper (2015). This paper elaborates on identifying the parameters to strengthen physical layer security. To overcome these challenges, the paper has highlighted the opportunities for applying physical layer security to the real systems. The emphasize is on physical layer security as the best approach for securing many emerging wireless systems, where conventional cryptographic approaches are not suitable due to resource constraint nature of devices. As most of the energy is consumed for core functions, there may be little left over for supporting security. To address these low power issues, the author proposed to find the solution to address security at the physical layer with a practical impact on real systems.

Esraa Ghourab et al. (2017) considered IoT devices in the wireless network that interacts and cooperates with each other using the static channel for wireless transmission. This network is vulnerable to eavesdroppers. The work has conducted security evaluation based on optimal relay selection protocols to maximize the channel secrecy capacity and improve data transmission. This approach uses a channel sensing metric to manipulate the transmission characteristics between legitimate users via changing the operating frequency and bandwidth. As a result, a massive improvement in physical layer security against eavesdropping attack is observed, when using reconfigurable frequency bands.

Lin Hu et al. (2018) proposed a work to secure downlink transmission from a controller to an actuator, with the help of a cooperative jammer to fight against multiple passive and noncolluding eavesdroppers. In addition to artificial noise aided secrecy beam forming for secure transmission, Cooperative Jamming (CJ) is explored to further enhance physical layer security. The paper has provided secrecy enhancing transmit design to minimize the secrecy outage probability (SOP), subject to a minimum requirement on the secrecy rate. As a result of the mathematical analysis, they have characterized the impacts of the main channel quality and the minimum secrecy rate on transmit designs. The numerical results confirmed the enhancement in security and power efficiency as compared with the approach without CJ.

Nozaki & Yoshikawa (2018) presented a 'Shuffling Based Side-Channel Countermeasure for Energy Harvester', that has validated the solution to address side-channel analysis attack. In the proposed countermeasure, the author's randomized instruction execution cycles of each function F corresponding to S-BOX, which is a non-linear function of TWINE. The experimental results show that the processing time increased about 1.25 times compared with the conventional method.

Arvind Singh et al. (2018) addressed the key challenge of designing an ultra-lightweight, secured encryption engine to secure IoT edge devices from side-channel attack. For an ultra-low power image sensor node, authors explored the system level design space to propose an optimized data path architecture for 128-SIMON, a lightweight block cipher for minimal performance, power and chip area overheads with better side-channel security.

Junqing Zhang et al. (2016) used a test-bed constructed by using wireless open-access research platform, to perform a thorough experimental study on key generation principles. The principles include temporal variation, channel reciprocity, and spatial de-correlation. The authors did the three-step comprehensive study i.e. 1) carrying out several experiments in different multipath environments; 2) considering static, an object moving, and mobile scenarios in these environments, which represents different levels of channel dynamicity; and 3) studying two most popular channel parameters, i.e., channel state information and received signal strength. As a result of more than a hundred tests, the authors offered insights and guidelines for the key generation system design. In their finding, they have shown that multipath is essential and beneficial to key generation as it increases the channel randomness. The authors also found that the movement of users/objects could help to introduce temporal variation/randomness and help users reach an agreement on the keys.

Jongyeop Kim et al. (2018) considered smart eavesdropping attacks on Multiple-Input-Multiple-Output (MIMO) wiretap channels for a legitimate transceiver. The author presented a smart eavesdropper model and a cooperative jamming solution between transceivers. The solution can control the jamming signal power to achieve the optimal secrecy performance. In their proposed solution, for practical applications, they have considered the residual self-interference from the full-duplex receiver and the limited cancellation capability of the smart eavesdropper. The authors have investigated the impact of passive and smart eavesdropping attacks in the MIMO wiretap scenario for physical layer security. As a result, the proposed solutions can improve the secrecy performance significantly by exploiting the full-duplex receiver and the cooperative jamming strategies with the sophisticated power control according to power expenditure.

Kimbahune et al. (2018) proposed the key management scheme to protect Next Generation Internet from a replay attack. The proposed scheme has been evaluated with a performance metric like delay (transmission delay, processing delay, and propagation delay), in two scenarios, viz. centralized and decentralized networks.

Xingyu Chen et al. (2018) proposed a solution to combat cloning attack based on COTS RFID devices and the universal C1G2 standard, without any software redesign or hardware augment needed. The authors used the RF signal profile to characterize each tag. In their proposed solution, a clustering-based scheme is used to detect the cloning attack static scenario and a chain-based scheme for clone detection in the dynamic scenario.

Table 9.1 Gap analysis of solutions on physical layer attacks

Sr. No	Reference Paper	Year of Publication	A1	A2	A3	A4	A5	A6	A7
1	[17]	2015	–	–	–	–	–	√	–
2	[18]	2016	–	–	–	–	√	–	–
3	[19]	2016	–	–	–	–	√	–	–
4	[20]	2017	–	–	–	√	–	–	–
5	[21]	2017	–	√	–	–	–	–	–
6	[22]	2017	√	√	–	√	√	–	–
7	[24]	2017	–	–	–	–	√	√	–
8	[26]	2017	–	–	√	–	–	–	–
9	[29]	2017	–	–	–	–	√	–	–
10	[30]	2018	–	–	–	–	√	–	–
11	[31]	2018	–	–	–	–	–	–	√
12	[32]	2018	–	–	–	–	–	–	√
13	[34]	2018	–	–	–	–	√	–	–
14	[35]	2018	–	√	–	–	–	–	–
15	[36]	2018	–	–	√	–	–	–	–

A1: Denial of service, A2: Message replay, A3: Tag cloning, A4: Spoofing, A5: Eavesdropping,
A6: Jamming, A7: Side-channel.

The experimental results of the proposed schemes show the detection accuracy of their approaches reaches 99.8% in a static scene and 99.3% in a dynamic scene.

9.4 Gap Analysis

This section describes the analysis of solutions to handle physical layer attacks in IoT. The several researchers provided solutions to protect the devices, are represented in Table 9.1.

The analysis of the existing work represents the physical layer attack is less focused and work that is more effective is needed to address the physical layer attack.

9.5 Threat Overview and Activity Modeling of Attacks

Before we present the solution for physical layer attack, this section, presents important terms and their meaning. The terms are vulnerability, threat and risk are discussed. This section also presents a detailed description of specific threats to the physical layer of IoT architecture.

9.5.1 Vulnerability vs. Threat vs. Attack

Below are the definitions of these terms, in the context of IoT.

Vulnerability term refers to a flaw in a system that can leave it open to attack. This may also be referred to any type of weakness in a system itself or in a set of procedures, or in anything that leaves information security exposed to a threat. Minimizing vulnerabilities gives fewer options for malicious users to gain access to the system (see Technopedia).

A threat refers to a thing that has the potential to cause serious harm to a system and can lead to attacks. A threat is something that may or may not happen but has the potential to cause serious damage. Threats are potentials for vulnerabilities to turn into attacks on systems, networks, and more that leads to humungous risk to individuals' computer systems and business computers. Therefore, vulnerabilities have to be fixed so that attackers cannot creep into the system and cause damage (see Technopedia).

An attack is an information security threat that involves an attempt to obtain, alter, destroy, remove, implant or reveal information without authorized access or permission. There are different types of attacks like Passive, active, targeted, clickjacking, brandjacking, botnet, phishing, spamming, inside, outside etc. which may happen to both individuals and organizations(see Technopedia).

9.5.2 IoT Reference Model

The IoT is a collection of billions of heterogeneous devices to collect information from the environment around the device and connected with other devices via the internet. Figure 9.4, shows the four-layer IoT reference model (Yang et al., 2013).

Perception layer consists of a large collection of heterogeneous sensing devices and the actuators referred to as "Things" in an IoT scenario. Network layer includes all hardware and software entities needed for communication networks. The support layer includes all middleware technologies, used to implement IoT services and integrated services and applications. The application layer provides global management for applications based on information processed in the support layer.

Looking at the perception layer, RFID communication and Wireless Sensor Network (WSN) is the most commonly used technologies in IoT. RFID communication comprises four layers (Mitrokotsa et al., 2010) viz. The physical layer, Network-Transport layer, Application layer and Strategic layer, whereas WSN comprises five layers (Akyildiz et al., 2002) viz.

Figure 9.4 Four layer IoT reference model.
Source: Yang et al., 2013.

The physical layer, Data Link layer, Network layer, Transport layer and Application layer. Considering the scope of this paper, we are focusing on the Physical layer. In RFID communications, the physical layer encompasses the physical interface, the radio signals used and the RFID devices (Mitrokotsa et al., 2010). On the other side, in WSN, this layer is responsible for the selection of frequency, generation of a carrier frequency, signal detection, modulation & data encryption. In general, physical layer consists of sensors, actuators, computational hardware, identification and addressing of the things, with the purpose to recognize the data from the environment. This layer is also responsible for frequency selection, modulation-demodulation, encryption-decryption, transmission and reception of data. Therefore, data sensing, data collection and secured data transmission are the main functions of this layer.

9.5.3 Physical Layer Threats

There are different types of unique threats as well as common threats at several layers. Same threats may be major in one layer at the same time it may be minor in another layer. This section described the major threats in the physical layer.

- **Denial of Service (DoS):**

Denial of Service attacks are threats to all modern communication systems, which causes the system to work inefficiently.

Specifically, the DoS attack is aimed at distracting communication between RFID tags and readers by means of multiple tags or specially designed tags to exceed the reader's capacity to handle concurrent requests

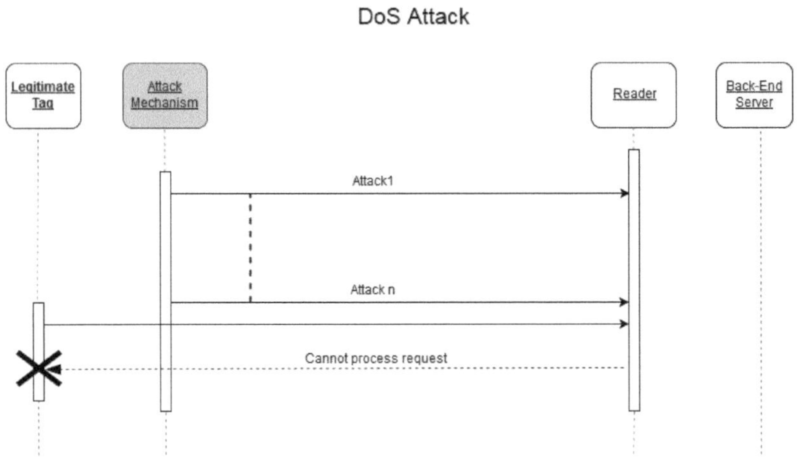

Figure 9.5   Denial of service attack.

(Spruit & Wester, 2013). Due to this attack, the system becomes inoperative, since the reader is unable to differentiate different tags, and the legitimate tags are unable to communicate with the reader. DoS attack in the scenario is presented in Figure 9.5. Here, the reader is having the capacity to handle 'n' concurrent requests.

Unauthorized tag disabling is also a type of DoS attack in RFID communication. This attacker causes RFID tags to a state where they can no longer function properly. This attack may lead to making tag inoperable temporarily or permanently. This type of attack is a serious threat to the integrity of the automated inventory management system.

- **Message Replay**

It is one of the most serious threats, which RFID communication face.

The replay attack occurs when a malicious node or device records a message and replay later. The replay attack is when an attacker replays that key information which is eavesdropped through the communication between reader and tag, in order to achieve cheating. Message replay may confuse the receiving device if it does not check for redundant messages. The modelling of this attack is presented in Figure 9.6.

- **Tag cloning**

Due to the availability of writable and reprogrammable inexpensive tags, it becomes very easy to clone the tags and use them without any expertise.

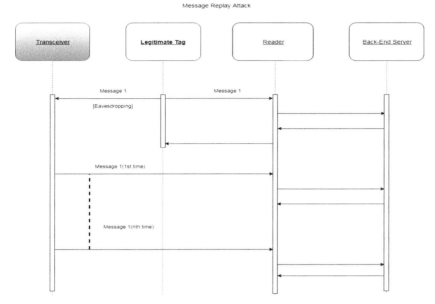

Figure 9.6 Message replay attack.

The tags without security features can easily be cloned by simply copying tag ID and associated data to the clone tag. The modelling of this attack is presented in Figure 9.7. The more sophisticated attack is required if the tag has extra security features to fool the reader to accept cloned tag as a legitimate one. Due to cloning, identical tags are circulated which creates confusion concerning the associated tagged objects and violates the integrity of the system.

- **Spoofing**

It is a variant of cloning that does not physically replicate an RFID tag. It occurs when an adversary impersonates a valid RFID tag to gain its privileges when communicating with an RFID reader.

Adversaries impersonate RFID tags using emulating devices with capabilities exceeding those of RFID tags, which include full access to the communication channel and complete knowledge of protocols and secrets required for authentication. Figure 9.8 depicts the spoofing attack and its modelling.

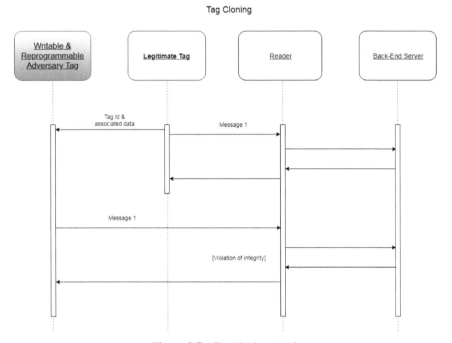

Figure 9.7 Tag cloning attack.

- **Eavesdropping**

In eavesdropping, adversary overhears and intercepts the data in the transmit coverage area of a node, without its knowledge. It is also known as skimming. With the radio receiving equipment, an attacker can listen, monitor, or record the communication between RFID tag and reader. In general, it is performed by reading the data, which is being broadcasted by the RFID system, which can be translated by the eavesdroppers into understandable information. It is very easy when the data is not protected and eventually allows other threats to take place. The attack details are represented in Figure 9.9.

- **Jamming**

Due to the nature of RFID tags to listen to all signals in its range, this attack happens in two ways, viz. passive interference and active jamming.

Passive interference happens due to unintentional sources like noisy electronic generators and power switching supplies. In addition to this, metal compounds, water or ferrite beads may also impair or even block the radio

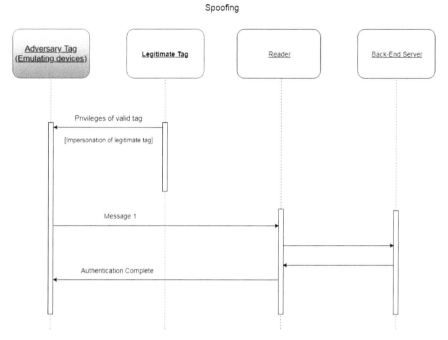

Figure 9.8   Spoofing attack.

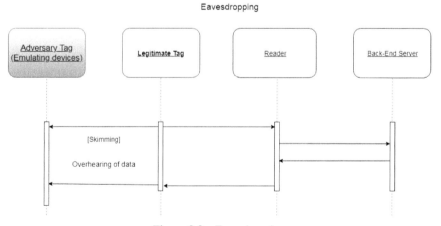

Figure 9.9   Eavesdropping.

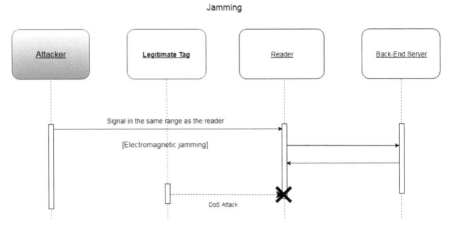

Figure 9.10   Jamming attack.

signal and lead to radio frequency detuning. Whereas in active jamming, an adversary may cause electromagnetic jamming by creating a signal in the same range as the reader in order to prevent tags to communicate with readers. It results in a DOS attack. Figure 9.10 illustrates a jamming attack.

- **Side-channel attack**

This attack takes advantage of the physical implementation of a cryptographic algorithm and information exploited is timing information, power consumption or even electromagnetic fields. To deploy a side channel attack requires deep knowledge of the host system on which cryptographic algorithms are implemented. Timing attacks can be implemented based on fluctuations in the rate of computation of the target while Simple Power Analysis (SPA) attacks extract information based on the variations of the power consumption. Differential Power Analysis (DPA) is a special type of power analysis attacks, which make use of the electromagnetic variations, produced during the communication between an RFID reader and tag to reveal secret cryptographic keys. The description of a side-channel attack is represented in Figure 9.11.

9.6 Issues and Challenges

The ubiquitous and pervasive nature of IoT has brought several benefits to the users, but due to the openness of IP network, it has become a soft target for attacks, which has been increased in the recent years. In addition to security,

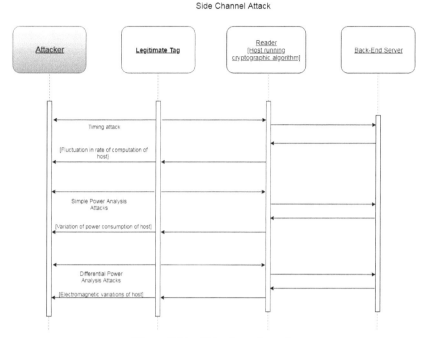

Figure 9.11 Side-channel attack.

interoperability and privacy are the other issues, which are also challenges for researchers.

Due to the additional cost involved and quick demands in the market, the manufacturers of IoT devices often neglect security requirements. Some devices manufactured with protection feature are usually provided software-based protection due to which hardware is left with some vulnerabilities, allowing new attacks on the system. As most of the IoT devices deployed do not have the option to add security mechanism by users, it becomes necessary to ship the products with built-in security.

Due to the heterogeneous nature of applications and devices, it has brought additional device specific attack surfaces, which have become more

Another most important issue is the IoT system composition. It consists of nodes with limited resources in terms of software, hardware, power, computational capabilities etc. (e.g. RFID or sensor nodes). So, it becomes necessary to design and develop a lightweight algorithm that can be used in these devices.

In the IoT system, the physical or perception layer is the most complicated layer to be protected because of the technological heterogeneity and the perceptual environment.

9.7 Proposed Methodology

Considering various issues and challenges associated with security at the physical layer, we propose the methodology and it is represented in the Figure 9.12.

This proposal defines, develops, implements/simulates and analyzes a novel lightweight algorithm to provide embedded security at the physical layer in M2M communications. The proposed methodology starts with defining IoT devices and their setup. Once the setup is ready, the next step is designing the security model. This algorithm consumes less energy and it is attack resistant. Once the algorithm is ready then the integration of sub-solutions is done to provide secure M2M communication.

To achieve the objectives, research work shall rigorously review the literature in the domain of lightweight embedded security for devices used in M2M communication. A detailed feasibility study will be done for each objective by reviewing the literature on each domain and presenting the understandings to the experts in that specific domain.

To summarize the methodology, Define – Measure – Analyze – Design – Verify (DMADV) implementation of Design for Six Sigma (DFSS) (Graves,

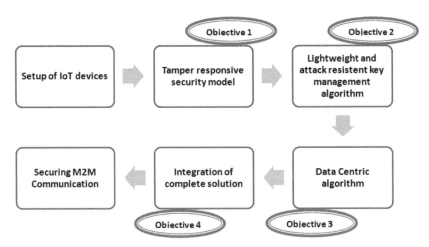

Figure 9.12 Proposed Methodology.

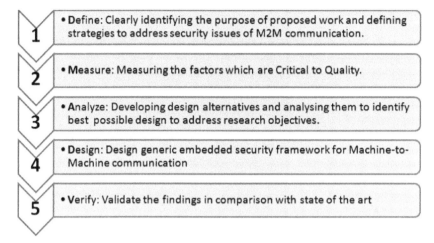

Figure 9.13 Proposed Methodology.

2002) shall be applied as a high-level research methodology, to address security issues in M2M communication towards IoT, as shown in Figure 9.13.

With its evolutionary approach, DMADV methodology is applied to existing products for the removal of defects to meet system requirements.

9.8 Conclusion and Future Outlook

This paper has explained IoT and its business case to explore the importance of IoT. The securing IoT devices are dramatically emerged due to the rapid development of IoT solutions. In this paper, we have seen different security issues associated with IoT, which need to be addressed. It is necessary to think appropriately to secure the IoT world by providing secure IoT communication protocols.

Based on the study, the physical layer of IoT is the most vulnerable layer due to the open nature of devices used, due to their constrained resources and due to technological heterogeneity.

Thus, in future, it becomes crucial to address the critical issues in IoT by implementing a lightweight algorithm to provide embedded security at the physical layer that can be adapted in heterogeneous environments with resource-constrained devices.

References

Akyildiz, I. F., Su, W., Sankarasubramaniam, Y., and Cayirci, E. (2002). Wireless sensor networks: a survey, vol. 38, pp. 393–422.

Babar, S. D. (2015). Security Framework and Jamming Detection for Internet of Things.

Babar, S., Stango, A., Prasad, N., Sen, J., and Prasad, R. (2011). Proposed Embedded Security Framework for Internet of Things (IoT), in *2nd International Conference on Wireless Communication, Vehicular Technology, Information Theory and Aerospace & Electronic Systems Technology (Wireless VITAE)*, p. 5.

Bi, Y. and Chen, H. (2016). Accumulate and Jam?: Towards Secure Communication via A Wireless-Powered, vol. 4553, no. c, pp. 1–13.

Cavoukian, A. and Reed, D. (2013). Big Privacy: Bridging Big Data and the Personal Data Ecosystem Through Privacy by Design.

Charalampos, M., Konstantinos, F., Alexandros, P., and Ioannis, P. (2014). Embedded systems security: A survey of EU research efforts. *Security and Communication Networks*, 8(11), 2016–2036.

Chen, X., Liu, J., Wang, X., Zhang, X., Wang, Y., and Chen, L. (2018). Combating Tag Cloning with COTS RFID Devices, *2018 15th Annu. IEEE Int. Conf. Sensing, Commun. Netw.*, pp. 1–9.

Das, K. (2017). Internet of Things and Implications in a Developing Economy," *Vivekanand International Foundation*. Available: https://www.vifindia.org/article/2017/october/03/internet-of-things-and-implications-in-a-developing-economy.

Definitions on Techopedia. Available: https://www.techopedia.com/.

El Mouaatamid, O., Lahmer, M., and Belkasmi, M. (2016). Internet of Things Security?: Layered classification of attacks and possible Countermeasures, *Electron. J. Inf. Technol.*, 9(9), pp. 66–80.

Esfahani, A., et al. (2017). A Lightweight Authentication Mechanism for M2M Communications in Industrial IoT Environment, *IEEE Internet Things J.*, vol. 4662, no. c, pp. 1–8.

Ghourab, E. M., Mansour, A., Azab, M., Rizk, M., and Mokhtar, A. (2017). Towards physical layer security in Internet of Things based on reconfigurable multiband diversification, *2017 8th IEEE Annu. Inf. Technol. Electron. Mob. Commun. Conf. IEMCON 2017*, pp. 446–450.

Graves, A. (2012). What is DMADV?, *Six Sigma Daily*. Available: https://www.sixsigmadaily.com/what-is-dmadv/.

Hu, L. et al. (2018). Cooperative Jamming for Physical Layer Security Enhancement in Internet of Things, *IEEE Internet Things J.*, 5(1), pp. 219–228.

Huang, J., Li, X., Xing, C. C., Wang, W., Hua, K. and Guo, S. (2017). DTD: A Novel Double-Track Approach to Clone Detection for RFID-Enabled Supply Chains, *IEEE Trans. Emerg. Top. Comput.*, 5(1), pp. 134–140.

Kanuparthi, A. (2013). Hardware and Embedded Security in the Context of Internet of Things, in *The First International Academic Workshop on Security, Privacy and Dependability for CyberVehicles*, pp. 61–65.

Kermani, M. M., Zhang, M., Raghunathan, A., and Jha, N. K. (2013). Emerging frontiers in embedded security, *Proc. IEEE Int. Conf. VLSI Des.*, pp. 203–208.

Kim, J., Kim, J., Lee, J., and Choi, J. P. (2018). Physical-Layer Security Against Smart Eavesdroppers: Exploiting Full-Duplex Receivers, *IEEE Access*, vol. 6, pp. 32945–32957.

Kimbahune, V. V., Deshpande, A. V., and Mahalle, P. N. (2018). A Novel Key Management Scheme for Next Generation Internet: An Attack Resistant and Scalable Approach, *Int. J. Inf. Syst. Model. Des.*, 9(1), pp. 92–121.

Mahalle, P. N., Anggorojati, B., Prasad, N. R., and Prasad, R. (2013a). Identity Establishment and Capability Based Access Control (IECAC) Scheme for Internet of Things, *J. Cyber Secur. Mobility, River Publ.*, 1(4), pp. 309–348.

Mahalle, P. N., Prasad, N. R., and Prasad, R. (2013b). Novel Context-aware Clustering with Hierarchical Addressing (CCHA) for The Internet of Things (IoT). Fifth International Conference on Advances in Recent Technologies in Communication and Computing (ARTCom 2013) 20–21 Sept. 2013.

Mahalle, P. N., Prasad, N. R., and Prasad, R. (2014). Threshold Cryptography-based Group Authentication (TCGA) Scheme for the Internet of Things (IoT), in *IEEE 3rd International Conference on Wireless Communications, Vehicular Technology, Information Theory and Aerospace & Electronic Systems Technology*.

Ministry of Urban Development Government of India, *Smart Cities: Mission Statement & Guidelines*, 2015.

Ministry of Urban Development Government of India, *Smart Cities Mission: Two Years of Fulfilling Promises: Short Film*. India, 2017.

Mitrokotsa, A., Rieback, M. R., and Tanenbaum, A. S. (2010). Classifying RFID attacks and defenses, *Inf. Syst. Front.*, 12(5), pp. 491–505.

Mukherjee, A. (2015). Physical-Layer Security in the Internet of Things: Sensing and Communication Confidentiality Under Resource Constraints, *Proc. IEEE*, 103(10), pp. 1747–1761.

Nozaki, Y. and Yoshikawa, M. (2018). Shuffling Based Side-Channel Countermeasure for Energy Harvester, in *2018 IEEE 7th Global Conference on Consumer Electronics (GCCE)*, pp. 714–715.

Pecorella, T., Brilli, L., and Mucchi, L. (2016). The role of physical layer security in IoT: A novel perspective, *Inf.*, 7(3), 49.

Rose, K., Eldridge, S., and Chapin, L. (2015). The Internet of Things: An Overview. Understanding the Issues and Challenges of a More Connected World, *Internet Soc.*, p. 80.

Shi, Z., Chen, J., Chen, S., and Ren, S. (2017). A lightweight RFID authentication protocol with confidentiality and anonymity, *Proc. 2017 IEEE 2nd Adv. Inf. Technol. Electron. Autom. Control Conf. IAEAC 2017*, pp. 1631–1634.

Singh, A., Chawla, N., Ko, J. H., Kar, M., and Mukhopadhyay, S. (2018). Energy Efficient and Side-Channel Secure Cryptographic Hardware for IoT-edge Nodes, *IEEE Internet Things J.*, vol. PP, no. c, p. 1.

Spruit, M. and Wester, W. (2013). RFID Security and Privacy?: Threats and Countermeasures.

Sudarshan, P., Abhishek, V., and Gunjan, G. (2017). The Internet of Things: Revolution in the making.

Taylor, S. and Boniface, M. (2017). Next Generation Internet the Emerging Research Challenges Key Issues Arising From Multiple Consultations.

Tomiæ, I. and McCann, J. A. (2017). A Survey of Potential Security Issues in Existing Wireless Sensor Network Protocols, *IEEE Internet Things J.*, 4(6), pp. 1910–1923.

Trappe, W. (2015). The challenges facing physical layer security, *IEEE Commun. Mag.*, 53(6), pp. 16–20.

Wang, N., Li, W., Jiang, T., and Lv, S. (2017). Physical layer spoofing detection based on sparse signal processing and fuzzy recognition, *IET Signal Process.*, 11(5), 640–646.

Yang, G., Xu, J., Chen, W., Qi, Z. and Wang, H. (2010). Security characteristic and technology in the Internet of Things, *J. Nanjing Univ. Posts Telecommun.*, 30(4), pp. 20–29.

Yi-Sheng, S., Yu, C. S., Hsiao-Chun, W., Huang, S. C. -H., and Hsiao-Hwa, C. (2011). Physical layer security in wireless networks: A tutorial, *IEEE Wirel. Commun.*, 18(2), pp. 66–74.

Zhang, J., Duong, T. Q., Woods, R., and Marshall, A. (2017). Securing Wireless Communications of the Internet of Things from the Physical Layer, An Overview, no. Llc, pp. 1–10.

Zhang, J., et al. (2016). Experimental Study on Key Generation for Physical Layer Security in Wireless Communications, *IEEE Access*, vol. 4, pp. 4464–4477.

Zhu, J., Zou, Y., and Zheng, B. (2017). Physical-Layer Security and Reliability Challenges for Industrial Wireless Sensor Networks, *IEEE Access*, vol. 5, no. c, pp. 5313–5320.

10

An Extension of the Information Systems Success Model; A Study of District Health Information Management System (DHIMS II) in Ghana

Patrick Ohemeng Gyaase[1] and Kodua Bright[2]

[1]Faculty of Information and Communication Science & Technology
Catholic University College of Ghana, Fiapre, Sunyani, Ghana
[2]Holy Family Hospital, Nkawkaw, Ghana
pkog@cug.edu.gh, brightkoduah@gmail.com

Health information system is credited with the capabilities to collect, process, report and use health information and knowledge to influence policy and decision-making. The District Health Information Management System (DHIMS II) is a comprehensive health information system for capturing, reporting and analysing health data for the healthcare ecosystem in Ghana. The system has been operational since 2012 and is available in all 216 health districts in Ghana. This paper evaluates this health information management system using an adaptation of Information Systems (IS) Success Model by Delone and McLean. The study introduces three implementation variables, namely, Management Support, Resource Availability and User Involvement, Education and Training to the IS success model to assess their influence on the health information system's success. Stratified sampling and simple random sampling were used for the selection of the districts from which quantitative data were collected using questionnaires. Purposive sampling was then used to select the respondents. The results shows the system quality of District Health Information Management Systems (DHIMS II) has improved information quality, service quality. These have significantly influenced the users' intention to continue using the system. The study also affirmed the significance of the impact that the management support, resource availability and user involvement, education and training has on the systems use and continued usage. The study concludes that although the implementation of the system has been successful, there are concerns over data integrity hence the need restrict the number users with permission to edit data in the system.

10.1 Introduction

Reliable information is the foundation of decision-making in healthcare delivery. It is also essential for healthcare policy development and implementation, governance and regulation, health research, human resources development, health education and training, service delivery and financing (Hodge, 2011). Health Information System (HIS) has therefore become such an important component of the healthcare system worldwide but not much attention has been given to it in most developing countries for years, affecting the quality of healthcare delivery (Wang et al., 2018).

According to Sharma et al. (2016), health information systems facilitate effective decision making at all levels of healthcare system by generating timely information, identify problems and needs, make evidence-based decisions on health policy and allocate scarce resources optimally. According to WHO (2015), an effective Health Information system that provides concise, relevant information gives donors and global partnerships a stronger foundation for their support of public health programs. Also, reliable and timely health information can mean the difference between life and death, most serious global health problems are in poorer countries, and good health information is vital for tackling them (Kasambara et al., 2017).

Ghana Health Service (GHS) thus adopted District Health Information Management Systems (DHIMS II), which is an open source, web-based health information system developed by University of Oslo, which could be implemented both online and offline originally referred to DHIS 2 (http://www.hisp.org, 2018). With the implementation of the system in Ghana, it has been re-christened DHIMS II. DHIMS II's modules includes data entry and capture, reporting and tracker module for data visualization (Nyonator et al., 2013).

Prior to the implementation of DHIMS II, Health related data were captured on spreadsheet application, which made the data management process prone to errors and lack of trust in information generated for decision making. Data processing was tedious and decision makers did not have access to timely analysed information for health decision making (Mensah, 2016).

This study, therefore evaluates the success of the district health information management system (DHIMS II) in meeting information needs of the healthcare delivery ecosystem in Ghana and its impact on Health service delivery since its adoption.

10.1.1 Quality Characteristics of Health Information System

The primary objective of Health information systems is to capture and provide quality data. Data quality possesses features and characteristics that ensures its ability provide timely information for decision making and planning (Debattista et al., 2016).

Various studies have highlighted various quality characteristics of data which could be compressed into five main attributes;

- **Access Security**: Data generated about entities must be restricted and kept secured and should be accessed by only authorized people to ensure confidentiality and privacy.
- **Accessibility:** Data should be available or easily retrievable for decision making as and when it is needed and for the purpose for which it was collected.
- **Accuracy:** Inaccurate data leads to wrong decisions thereby negatively affecting smooth delivery of health services. Therefore data captured must be correct and free of errors to enhance service quality.
- **Completeness:** Data captured must be complete, that is to have all indicators or fields appropriately filled, it must be sufficient in breadth, depth, and scope for its desired use.
- **Timeliness:** Data should be recorded as quickly as possible and used within a reasonable time period. Timeliness of data improves its usefulness in decision making. Outdated data mostly results in data losing its value (Batini & Scannapieco, 2016).

It is believe that poor and inaccurate data is hampering global aid efforts to improve the lives of the world's poor and quality health outcomes. It is therefore imperative to ensure that health information systems generate data and information that meet all the quality attributes as much as possible (Gotz & Borland, 2016).

10.1.2 Challenges in Health Information Systems Implementation

Implementation of health information systems is fraught with various challenges in spite of the huge potential to radically transform healthcare especially in developing countries. These challenges can be classified under technological, human, and organizational factors affecting the diffusion and utilization of health information systems in developing countries (Thangasamy et al., 2016).

Technologically, there is lack of standardization and interoperability of technology in the different health information systems used in the healthcare ecosystem in Ghana. There is not well-developed integrated healthcare information system which enables healthcare institutions to freely share healthcare data. Any attempt to implement such systems generate serious concerns on privacy and security of electronically transmitted health data (Mukasa et al., 2017).

From organizational perspectives, healthcare managers and policy makers are often concerned about the costs, return on investments. Health information systems are often perceived from the standpoint of the prevailing socio-political and economic system and conditions within which the technology is supposed to be embedded (Shah et al., 2016).

Among the human challenges are widespread resistance to diffusion and utilization of health information systems especially among clinicians in developing countries. Physicians and other healthcare personnel often have different perspectives of the system from that information technology (IT) personnel and healthcare managers (Ngafeeson, 2014). There is also low technical capacity or knowledge of workers on IT usage in various healthcare institutions in the country.

10.1.3 The Conceptual Model for the Study; an Extended Information Systems Success Model

In assessing the success of the District Health Information Management System (DHIMS II), this study utilizes an adaptation of the Information Systems (IS) Success Model (Delone & McLean, 2003). The IS Success model provides six comprehensive variables and their relationships in assessing information systems success, as seen in Figure 10.1. The Model postulates

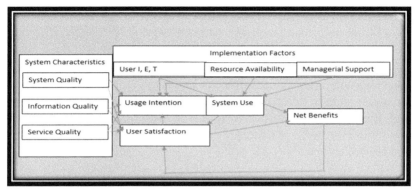

Figure 10.1 An Adaptation of IS Success Model.

Source: Delone & McLean, 2003.

that information quality, systems quality and service quality influence user intention and the actual use of an information system. These in turn influence individual user satisfaction and consequently influence the net benefit of the information system. This study, in addition to these variables, posits that the success of health information systems also depends on three additional implementation moderating factors. These are management support (Namakula & Kituyi, 2014), user involvement, education and training (Eden et al., 2016) and resource availability (Yi et al., 2008).

Information quality is a measure of the degree of quality of information stored, delivered or produced from an information system. The quality of information produced from an information system influences user's intention to continue using the system (Petter et al., 2008).

The System quality measures the extent to which the systems meets the purpose for which it was developed and is able to deliver the expected benefits. System quality leads to user satisfaction which in turn influence on actual use (DeLone & McLean, 2002).

Service quality measures the extent to which the information systems effectively deliver the services for which is developed. System Use and Usage Intentions are influenced by system characteristics namely, information quality, system quality, and service quality. System Use influences a user satisfaction with the information system, which, in turn, influence Usage Intentions. In addition to user satisfaction, system use directly affects net benefits that the system provides to the organization (DeLone & McLean, 2016).

User satisfaction measures of the extent to which users are pleased or contented with the information system, and is posited to be directly affected by system use, and by information, system, and service quality. Like actual system use, user satisfaction directly influences the net benefits provided by an information system (Delone & McLean, 2003).

Net systems benefit of an information system measures the extent to which a particular information system is able to deliver the overall value of the system to its users or to the implementing organization. In the IS success model, net system benefits are affected by system use and by user satisfaction with the system. In their own right, system benefits are posited to influence both user satisfaction and a user's intentions to use the system (DeLone & McLean, 2016).

10.2 Moderating Implementation Factors for the Success of Health Information Systems

Various studies including Namakula & Kituyi (2014) have all identified management support, resource availability and user involvement, education and training as influencing the success of information systems in organizations. This study thus postulates that these implementation moderating factors are also critical to the success of the district health information management in Ghana.

- Top management's unwavering commitment to health information systems implementation makes it an organization priority. Top-down approach has proven to be successful when an organization is considering the diffusion of technological innovation (Hwang, 2016). Top management support provides direction and e-leadership toward the implementation of information system thus influencing system use.
- User involvement in terms of education and training ensures that a shared vision for the information system being implemented especially if the goals and benefits of health information system clearly defined, meaningful and measurable to users. User education and training targeting the right group of users, with right message, at the right time is essential for successful deployment of health information systems especially for clinicians and other non-technical users. The training and education goes beyond the system's functionality, but should relate to the role of users and their appreciation of the benefits of the system (Yi et al., 2008). User involvement, training and education makes users

knowledgeable in the use of the systems and appreciate its value to the organization thus influences usage intention and system use.

- Availability of resources in terms of adequate finances, technical infrastructure, and other requisite resources ensure that the not only is implementation successful but continued utilization is ensured for the achievement of the desired goals. Resource availability thus, influences continued usage of the system, hence crucial toward the achievement of success in health information system (Gil-Garcia & Sayogo, 2016).

10.3 Methodology

This study utilized a non-interventional case study design using quantitative approach as methods of inquiry which involves the collection and analysis of quantitative data for the study (Creswell & Clark, 2007). Stratified random sampling, where the target population was divided into various strata and then random sampling was used to select districts from each region in Ghana (Black, 1999). This sampling technique was chosen to avoid sampling bias since it gives all districts, Municipalities and metropolitan area equal chances of being selected.

The study population comprises of all DHIMS II users, made up of health information officer, biostatisticians and other staff designated to capture data in the system. District directors and other decision makers who utilize data from the DHIMS II for decision and policy making were part of the study population. The study involved the distribution of questionnaire to 140 respondents were purposively sampled nationwide ranging from Health Information officers, Disease control officers, Nutrition officers, Biostatistics officers as well as metropolitan, municipal and district health directors. The results are presented in the form of charts and tables with frequencies and percentages for ease of understanding.

10.4 Data Analysis and Discussion of Results

A web-based structured questionnaires designed using Google forms with a five-point Likert scale were administered to 140 respondents of which 99 responded. The system was restricted to permit just a single response per an email account to lessen the tendency of multiplicity of responses from a user. Non responses were also curtailed by setting the system to ensure that all questions are answered before submission can be possible. These responses were monitored on daily basis until significant number of responses were

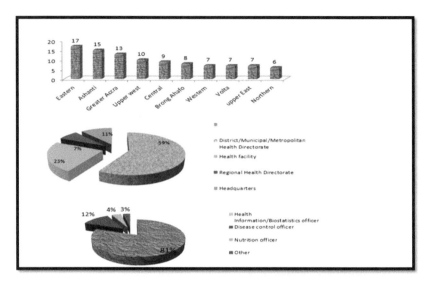

Figure 10.2   Distribution of Respondents.

received. The results are presented in the form of charts and tables with mean scores and standard deviation. The mean score measures the average of the scores of the variables whilst the standard deviation measures the extent of the dispersion of the responses from the mean. The validation of the extension of the IS Success model was done using inferential statistics, specifically regression analysis to show whether three identified implementation variables have an impact and the extent of the impact on the success of DHIMS II.

10.4.1 Demographic Data of Respondents

The responses were received from all the regions in the country. The highest responses came from the Eastern region and the least from the Northern region. The regional distribution of respondents is presented in Figure 10.2 below. Figure 10.2 further shows the distribution of respondents in terms place of work as well as the occupation for the respondents. It shows that 59% were from the district level followed by the health facility which accounted 23%. These are the level at which data is captured by the DHIMS II. 81% of the respondents were Health Information or Biostatistics officers whilst 12% were Disease Control Officers.

Table 10.1 System Quality of DHIMS II

System Quality	Mean	Std. Deviation
Variables		
Captures relevant indicators/performance	4.05	0.84
User friendly	4.26	0.63
Upgradeable /Modifiability	3.47	1.11
Availability	3.38	1.21
Processes data in real time	3.45	1.01
System Quality	**3.81**	**0.90**

Source: Field work.

10.4.2 An Analysis System Quality of DHIMS II

Achieving system quality must be considered throughout design, implementation, and deployment. No quality attribute is entirely dependent on design, nor is it entirely dependent on implementation or deployment. The system quality attributes of information System assessed are whether the systems is able to capture all the relevant data, usability measured by user friendliness, modifiability and performance. System quality of DHIMS II directly impacts the extent to which the system is able to deliver benefits by means of mediational relationships through the usage intentions and user satisfaction.

Table 10.1 above provides an analysis of users' perception of the systems quality of DHIMS II. The mean score of all the variables assessed indicates high system quality of DHIMS II. The mean score performance is 4.05 with 0.84 standard deviation. User friendliness has a mean score of 4.26 with 0.63 standard deviation, score, modifiability as measured by availability of upgrades has a mean score of 3.47 with standard deviation of 1.11 and processing of information in real time has a mean score of 3.45 and a standard deviation of 1.01. The overall mean score for system quality is 3.81 with standard deviation of 0.90. The responses shows that generally the users the system quality of the district health information management system as high. This has had positive influence on usage intention and user satisfaction of the system.

10.4.3 An Analysis of Information Quality of DHIMS II

Information quality of DHIMS II was measured by the extent to which the system is able to store and protect data, process data and deliver or produces reports to meet the needs of the users. This was measured using the information quality attributes such as data security, accessibility, reliability

Table 10.2   Information Quality of DHIMS II

	Mean	Std. Deviation
The Data Security	4.16	0.72
Ease of Data Retrieval/Accessibility	4.04	1.03
Data Reliability	4.00	0.64
Data Consistency	4.28	0.64
Generates Quality	4.26	0.60
Information Quality	**4.15**	**0.73**

Source: Field work.

and timeliness. Information quality impacts both a user's satisfaction with the system and the user's intentions to use the system, which, in turn, impact the extent to which the system provides net benefits for the user or the organization.

Table 10.2 provides an analysis of the users' perception of the information quality of DHIMS II. Data security has a mean score of 4.16 with 0.72 standard deviation, ease of data retrieval which measured by accessibility to data has mean score of 4.04 with standard deviation of 1.03, Data reliability has a mean score of 4.00 and a standard deviation 0.64, Data consistency has mean score of 4.28 and a standard deviation of 0.64 whilst generation of quality has a mean score of 4.26 and standard deviation of 0.60. These show high information quality from the system from users' perspectives. The overall information quality has a mean score of 4.15 with a standard deviation of 0.73. This would have positive influence on usage intention and the net benefits of the systems to health care delivery in Ghana.

10.4.4 An Analysis of Service Quality of DHIMS II

In assessing the service quality of DHIMS II, emphasis was placed on service delivery of the system, whether these service delivery meets the expectation of the user. The main purpose of the systems is to provide timely reports to decision makers and serving as the health data repository so donor partners can access data for the evaluation of their programmes. Service quality directly impacts on usage intentions and user satisfaction with the system, which, in turn, impact on the net benefits produced by the system to healthcare organizations.

Table 10.3 provides a descriptive analysis of the responses for service quality of DHIMS II. The analysis shows that the system ability to generate and transmit timely reports has a mean score of 4.53 with a standard deviation

Table 10.3   Service Quality of DHIMS II

Service Quality	Mean	Std. Deviation
Timely reports generation and transmission	4.53	0.52
Provides features for Data analysis	4.45	0.63
Effective source of Health service data	4.19	0.99
Effective in Managing Data	4.05	0.81
Service Quality	**4.31**	**0.74**

Source: Field work.

0.52, the provision of features for data analysis has a mean score of 4.45 with a standard deviation of 0.63, effective source of health service data for decision making has a mean score of 4.19 and a standard deviation of 0.99 whilst its effectiveness in managing health data has a mean score of 4.05 and a standard deviation of 0.81. These results show high service quality of the systems. The overall service quality has a mean score of 4.31 with a standard deviation of 0.74. This show that the users perceive the system as having high service quality. This will have positive influence on usage intention, actual use and net benefit to the health service delivery ecosystem in the country.

10.4.5 The Impact of Implementation Factors on the Success of DHIMS II

The conceptual model for this study was and extension of the information systems success model. Three additional implementation factor were proposed as variables contributing significantly to the success of the DHIMS II. These are management support, user involvement, education and training and resource availability. Assess the impact of these variables, regression analysis used to determine the significance of their contribution to the success of the system. The criterion variable is the variable (DHIMS II success) and the predictor variables are management support, user involvement, education and training and resource availability. This was to determine the dichotomy among the criterion and predictor variables.

The regression analysis resulted in the R and the R square value being the same at (1.00) which shows a perfect positive correlation between the predictor variables management support, user involvement, education and training and resource availability and the Criterion variable (DHIMS II success). This shows a strong positive relationship between these variables and the success of DHIMS II.

Table 10.4 Coefficients of independent variables

Model	Unstandardized Coefficients		Standardized Coefficients			95% Confidence Interval for B	
	B	Std. Error	Beta	t	Sig.	Lower Bound	Upper Bound
(Constant)	3.764E-14	.000		.000	1.000	.000	.000
Management support	1.000	.000	.394	1.430E8	.000	1.000	1.000
User involvement education and training	1.000	.000	.425	1.620E8	.000	1.000	1.000
Resource availability	1.000	.000	.476	1.711E8	.000	1.000	1.000

A Dependent Variable: DHIMS Success.
Source: Field work.

The Table 10.4 above presents the results of the linear regression analysis of the data from the respondent on the impact of the three proposed variables on the success of the DHIMS II implementation. The dependent variable was DHIMS II Success and the independent variables are management support, user involvement, education and training and resource availability. The results shows standardized coefficients (Beta) value for management support is 0.394, the beta value for User Involvement, education and training is 0.425 and resource availability has regression coefficient of 0.476 and 95% confidence interval.

The results show significant positive impact of the proposed variables on the success of the DHIMS II implementation in Ghana. It can therefore be concluded that the success achieved with the implementation of DHIMS II has been made possible not only as results of the system, information and service qualities of the system, but management support, user involvement, education and training and resource availability contributed significantly to the perceived success from the users perspectives.

10.5 Findings

The findings were drawn from the outcome of the analysis of the responses from respondents based on the various constructs used to measure the success of the DHIMS II. The system quality, Information quality and service quality of DHIMS II were assess from the users' perspective using descriptive statistics.

Regression analysis also carried test to validate the proposed additional variables and their impact on success of DHIMS II implementation thereby expanding the constructs of IS Success model by Delone and McLean to include management support, user involvement, education and training and availability of resources.

10.5.1 System Quality, Information Quality and Service Quality of DHIMS II

Findings from the analysis of data from the respondents shows that users perceive DHIMS II has high systems quality. All the systems variable measured have high mean score with considerable low standard deviation indication widespread agreement on the quality of the system. The respondents' perspective of DHIMS II its implementation satisfies system quality characteristics. This therefore has had positive influence of user intention and actual use as respondents continue to use the system. This has provided net benefit for Ghana health service as it continues to support the operation of the systems and investing in infrastructure to expand its coverage of instructions and the donors continue to support the implementation of the system.

The findings from the analysis also revealed DHIMS II software since its implementation has high information quality. All the variables for information quality assess have high mean score and considerable low standard deviation showing wide spread agreement on the information quality of DHIMS II indicating that information from the system can be relied on for decision making. These have contributed to the positively to user intention and actual use of the system resulting in net benefit for health service delivery. Detailed analysis of the responses showed users concern on the batch processing of data in areas where there is no internet coverage.

The study also found that users perceive the DHIMS II as having high service quality, thereby meeting users' expectation. Again all the variable of service quality assess showed high mean score with low corresponding standard deviations indication widespread agreement of respondents on the service quality of the system. This has resulted in increased donor support for the implementation of the system, continues use and net benefit to health service delivery in Ghana.

10.5.2 The Impact of the Extended Variable on DHIMS II Success

The regression analysis of the data from the responses for the study shows significant impact that management support, user involvement, education and training as well as resource availability have had on the success of DHIMS II. Thus in addition to standard predictor variables of the information systems success models namely, Service Quality, Information Quality, System Quality, the three proposed variables contributes to successful implementation of health information systems. These findings also validate (Hwang, 2016) and (Gil-Garcia & Sayogo, 2016).

10.6 Conclusions

This study was carried out to determine the factors that has contributed to the success or otherwise of the DHIMS II project in Ghana. The study identified the Information systems success model as the conceptual framework. However, a survey of literature showed three additional variables that were hypothesized as having contributed to the success of otherwise of the system, hence and extended IS model. The model has been validated indicating that also the IS success model is still a valid tool for the analysis of information systems success, demand of institution and stakeholders should be factored into the implementation of and information systems to ensure positive user intention and actual use of the system. User resistance, lack of management support and starving s new systems' implementation of resources would definitely leads to implementation failure no matter how much the systems meets the quality characteristics as postulated in the original IS success model. And this has been clear amplified by this study.

References

Batini, C. and Scannapieco, M. (2016). *Data and Information Quality: Dimensions, Principles and Techniques*. Springer.

Black, T. R. (1999). *Doing Quantitative Research in the Social Sciences: An Integrated Approach to Research Design, Measurement and Statistics*. London: Sage.

Creswell, J. W. and Clark, P. V. (2007). *Designing and conducting mixed methods research*: Wiley.

Debattista, J., Sören, A., and Christoph, L. (2016). Luzzu–A Framework for Linked Data Quality Assessment. *IEEE Tenth International Conference on Semantic Computing (ICSC)*. IEEE.

DeLone, W. H. and McLean, E. R. (2002). Information systems success revisited. System Sciences, *HICSS. Proceedings of the 35th Annual Hawaii International Conference* (pp. 2966–2976). IEEE.

Delone, W. H. and McLean, E. R. (2003). The DeLone and McLean model of information systems success: a ten-year update. *Journal of management information systems*, 19(4), 9–30.

Delone, W. H., & McLean, E. R. (2016). Information Systems Success Measurement. Foundations and Trends§in *Information Systems, 2(1)*, 1–116.

Eden, R., Fielt, E., and Murphy, G. D. (2016). The impact of user capital on information systems success. Pacific Asia Conference on Information Systems (PACIS). *PACIS 2016 Proceedings*. Retrieved from http://aisel.aisnet.org/pacis2016/400

Gil-Garcia, J. R. and Sayogo, D. S. (2016). Government inter-organizational information sharing initiatives: Understanding the main determinants of success. *Government Information Quarterly*, 33(3), 572–582.

Gotz, D. and Borland, D. (2016). Data-driven healthcare: Challenges and opportunities for interactive visualization. *IEEE computer graphics and applications*, *36*(3), 90–96.

Hodge, S. N. (2011). *Health Information System (HIS)*. Pacific health Information Network.

http://www.hisp.org. (2018). http://www.hisp.org/services/dhis-2/. Retrieved from http://www.hisp.org.

Hwang, M. (2016). Top Management Support, Task Interdependence and Information Systems Implementation Success: The Role of Common Method Variance and Measurement Error. *Communications of the ICISA*, 17(1), 30–47.

Kasambara, S., et al. (2017). Assessment of implementation of the health management information system at the district level in southern Malawi. *Malawi Medical Journal,* 29(3), 240–246.

Mensah, M. (2016). Management of records in state-owned hospitals in Ghana. *International Journal of Business Excellence*, 9(4), 463–487.

Mukasa, E., Kimaro, H., Kiwanuka, A., and Igira, F. (2017). Challenges and Strategies for Standardizing Information Systems for Integrated TB/HIV Services in Tanzania: A Case Study of Kinondoni Municipality. *The*

Electronic Journal of Information Systems in Developing Countries, 79(1), 1–11.

Namakula, S. and Kituyi, G. (2014). Examining health information systems success factors in Uganda's Healthcare System. *The Journal of Global Health Care Systems*, 4(1).

Ngafeeson, M. (2014). *Healthcare Information Systems: Opportunities*. In M. Ngafeeson, Healthcare Information Systems: Opportunities (pp. 261–262).

Nyonator, F., Ofosu, A., and Osei, D. (2013). District Health Information Management System DHIMS II: The Data Challenge For Ghana Health Service. Accra: Policy Planning Monitoring and Evaluation Division, Ghana Health Service.

Petter, S., DeLone, W., and McLean, E. (2008). Measuring information systems success: models, dimensions, measures, and interrelationships. *European Journal of Information Systems*, 17(3), 236–263.

Shah, G. H., Vest, J. R., Lovelace, K., and Mac McCullough, J. (2016). Local health departments' partners and challenges in electronic exchange of health information. *Journal of Public Health Management and Practice*, 22(Suppl 6), S44.

Sharma, L., Chandrasekaran, A., Boyer, K. K., and McDermott, C. M. (2016). The impact of health information technology bundles on hospital performance: An econometric study. *Journal of Operations Management*, 41, 25–41.

Thangasamy, P., Gebremichael, M., Kebede, M., Sileshi, M., Elias, N., and Tesfaye, B. (2016). A Pilot Study on District Health Information Software 2: Challenges and Lessons Learned in A Developing Country: an Experience From Ethiopia. *International Research Journal of Engineering and Technology*, 3(5).

Wang, Y., Kung, L., and Byrd, T. A. (2018). Big data analytics: Understanding its capabilities and potential benefits for healthcare organizations. *Technological Forecasting and Social Change*, 126, 3–13.

WHO (2015). The need for strong health information systems. Health Metrics Network.

Yi, Q. et al. (2008). Integrating open-source technologies to build low-cost information systems for improved access to public health data. *International Journal of Health Geographics*, 7(1), 29.

11

Reviewer Paper Assignment Problem – A Brief Review

Aboli H. Patil and Parikshit N. Mahalle

Department of Computer Engineering,
Sinhagad College of Engineering, Pune, India
aboleee.patil@gmail.com, alborg.pnm@gmail.com

Since last two decades, conferences are flooded with papers receiving excessively high numbers of submissions. Assignment of submitted papers to appropriate expert reviewers, known as Reviewer Assignment Problem (RAP), is a crucial and challenging task for journal editors and conference management committee. Core objective of RAP solution is fair and timely review process. A brief review of the literature is presented in this paper covering a typical conference management system, popular techniques, performance measures, data sets and research opportunities along with challenges in domain. Adaptation of 5G and use of machine learning algorithms will contribute to efficient conference management system.

11.1 Introduction

We are witnessing the era in which number of research publications and conferences are exponentially growing. One of the important tasks is getting proposals and papers reviewed fulfilling deadlines with fair and balance assignment. Reviewer Assignment Problem (RAP) is defined as process of allotting reviewers to submitted papers. This process is very vital and is a complex task for editors of journal, chair of conference, and for research councils. Person with experience as member review committee (for research that proposals submitted at funding agency) or one who has served as a Technical Program Committee (TPC) chair for a conference knows reviewing is very much time consuming and complex task. Neural Information Processing Systems (NIPS) is one of the top tier annual international conferences on Machine learning. NIPS 2016 witnessed 2,400 paper submissions, 3,000 reviewers, and 8,000 attendees (Shah et al., 2017).

Looking at number of conferences scheduled in different categories, it is revealed that the count is huge and it is increasing every year. As per the allconferences.com as indicated in Figure 11.1 the count of conferences scheduled in span of November 2018 to December 2019 for different categories all together is 91681. This count is as per the registered conferences at allconferences.com. It is very clear that beyond this count there must be many conferences that are registered at many similar service providers. Business, Computers, Health and Science are the major contributing categories.

The countries in Asia are leading in organizing conferences. Figure 11.2 indicates contribution for major countries in Asia for span of November 2018 to December 2019, as extracted from conferencealerts.com. Further it is observed that major contributing countries include India, China, Thailand, Singapore, Japan and UAE.

The process of RAP consists of two core steps- compute the match among reviewers and papers, and define process of reviewer assignment so as to

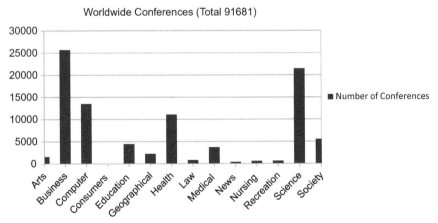

Figure 11.1   Current status count of Conferences scheduled worldwide for span of Nov 2018 to Dec 2019.

Source: Allconferences.com.

Conferences in Major Countries of Asia (Nov 2018 to Dec 2019)

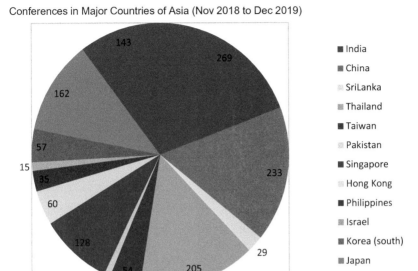

Figure 11.2 Current status of contribution of countries in Asia in scheduling Conferences (Nov 2018 to Dec 2019).

Source: conferencealerts.com.

optimize the assignment. Prime objective is to maximize the match satisfying most of the constraints within the feasible restrictions. Traditionally in semiautomatic systems, this task of reviewer assignment is handled by team of single or few persons so as to fulfill four major constraints: Good match between paper and reviewer's expertise, number of reviews per paper, reviewers' load balance and avoiding conflicts of interest (CoI). For fair review it is important to consider CoI while assigning reviewer to paper. CoI is a situation in which a person or organization is involved in multiple interests and serving one interest could involve working against another.

Reviewer assignment problem is typical assignment problem (Gani, & Mohamed, 2015) (Karimzadehgan et al., 2018). The assignment problem is a fundamental combinatorial optimization problem; solution aiming at finding a maximum weight matching or minimum weight perfect matching. Along with RAP, assignment problem has various practical applications as listed in Table 11.1. These variety application areas of real-world have motivated scholars in developing solutions for problem. RAP among wide range of applications, model their problem as assignment problem with objective to find optimal solution.

Table 11.1   Typical Assignment Problems

Reviewer-Paper Assignment in Conference Management
Resource allocation-matching funding agencies to research projects
Scheduling on parallel machines
Assigning managers to construction projects
Course-Teacher Assignment
Course-Examiner Assignment
Staff scheduling (assigning graduating students to interviewer)
Assigning press releases to newspaper reporters
Assigning staff to projects at consulting companies
Crew scheduling in airline companies
Posting Military Servicemen
Construction decision making or recommendation for material management
Educational Institute Time table Management
Patient-Doctor Assignment

Noticeable number of literature is available documenting contribution of various researchers in the domain. This paper is intended to provide a comprehensive survey so as to know what has been achieved, to identify gaps and to set direction for future research. This paper is organized as follows. Section 11.2 introduces process of conference management system in general. In Section 11.3 covers brief report of literature reviewed. Section 11.4 presents list of performance metric and data set used. Finally, Section 11.5 summarizes the survey listing research gap and challenges.

11.2 An Overview of Existing Conference Management Systems

Since last decade, conferences use web-based management system. Popular Conference Management Systems include – CyberChair, ConfMan, HotCRP, Softconf, Linklings, Websubrev, AAA S/W, Puma, SIGACT, ConfBits, WCMS, EDAS, OpenConf, INFOCOM, CMT, easy chair, ConfTool, mEngage, vCongress among many similar.

Such systems are semi-automatic or fully automatic leading to comparatively easier task handling. Facilities typically provided by such systems are like – paper submissions, collecting and handling of reviewers preferences, assigning reviewers, submission of reviews, download of papers by Program Committee (PC), review progress tracking, conduction of distant meeting, notification of review results; sending e-mails for notifications, uploading camera ready published papers and many related tasks. Though most of these tasks are supported by system, the crucial and time consuming task is assigning reviewers to manuscripts as this is not yet fully automated (Jin et al., 2017) (Karimzadehgan et al., 2008).

In Manual and semi-automatic method, paper assignment is often an obscure process, often done randomly and evenly among reviewers; at most, reviewers and papers are divided by tracks, corresponding to broad topics, and paper are distributed within each track (Jan & Contreras, 2015) (Mehlhorn et al., 2010).

Though said to be fully-automatic, in most of automatic systems, except upload and downloading of manuscripts, members of Program Committee Chair (PCC) of the conference complete this task by referring to the topic preferences, conflicts of interest and expertise domain of registered reviewers (Shah et al., 2017) (Kolasa & Krol, 2011). Many constraints are to be satisfied while assigning reviewers. Reviewers are assigned to papers either automatically or manually and if required the process is repeated. Depending on whether to be reviewed as double-blind or single-blind and based on single-level or two-level review structure; the reviewer assignment is done (Yarowsky & Florian, 1999).

Due to the many constraints to be fulfilled, the task is very tedious and difficult, and sometimes does not result in the best solution. When there is high count of papers and goal to satisfy constraints, it is noticed that current software provides fail to support for automatic review assignment. Often these conference management software just assist experts selecting domains to review and committee to refer these preferences (Dumais & Nielsen, 1992) (Wang et al., 2014).

Generally, the review process ends with the Program Committee meeting, where the papers are discussed on the basis of collected review forms, in order to accept or reject. Reviewer's comments are conveyed authors to improve if accepted or justification of rejection. In bidding process, conference management system asks the referees to declare conflicts of interests. It is expected from reviewers to rank the papers and conflict of interest as high, medium and low (Shah et al., 2017) (Rodriguez et al., 2007).

11.3 Related Work

The conference management and reviewer assignment problem are not very new domains, Dumais et al. were pioneering researchers to contribute for automating the assignment of submitted manuscripts to reviewers in 1992. Since then different dimensions of the reviewer-paper assignment problem have been studied by many researchers and is available in various forms. Selected papers brief study is presented with respect to process, technique, data sets, performance metric and pros-cons. The main constraints of RAP that are to be satisfied are – *Conflict of Interest* (CoI) – avoiding conflicts of interest between the reviewer and authors, *Authority* – expertise/interest of reviewer for paper domain, *Coverage* – Each paper should be reviewed a sufficient number of times and *Load Balance* – not to exceed maximum number of papers the reviewer wants review (Li & Hou, 2016) (Hartvigsen et al., 1999) (Wang et al., 2014).

Researchers Dumais et al. claimed in 1992 that it is possible to auto-mate the assignment of submitted manuscripts to relevant reviewers using information retrieval methods. Reviewers were identified for assigning papers using a variety of automated methods that based on information retrieval principles and Latent Semantic Indexing (LSI). 117 manuscripts submitted at the conference Hypertext-91 were successfully assigned among 15 reviewers (Dumais, & Nielsen, 1992). Technique computes the similarity between submitted papers and biographies of reviewers. Matrix is used to represent correlation between manuscripts and expertise/interest of reviewers and the dot product of the two matrixes compute matching degree. Then, assignment was done by picking several reviewers from those with high matching degree. These 15 reviewers provided ratings based on matching of abstract and their interests. And it is claimed that the automated process works fairly good as compared to manual method. A similar work is presented by researchers (see Yarowsky & Florian, 1999). The researcher has used published literature of reviewers instead of their biographies. The publication knowledge is extracted through information submitted by reviewers or with the help of web.

Merelo et al. contribution is process that combines greedy and an evo-lutionary approach for reviewer assignment. Author claims a good match between reviewers and papers assigned to them. The evolutionary algorithm helps to maximize match between the interest of reviewer and the domain of paper. The constraints satisfied are coverage and load balancing taking into account also incompatibilities and conflicts of interest (Merelo-Guervós & Castillo-Valdivieso, 2004). Researchers Fan Wang et al. presented the tech-nique that initially forms groups of manuscripts and reviewers and then groups of reviewers are assigned to groups of manuscripts. RAP problem is formulated as a multi-objective mixed integer programming model and the problem is NP-hard. An effective two-phase stochastic-biased greedy algorithm is designed to solve the problem. Researchers have tested approach for a real application, collecting feedbacks that were positive and encouraging (Wang et al., 2013). Authors A. Nagoor Gani et al. has developed a novel algorithm for solving assignment problems with costs as generalized trapezoidal intuitionistic fuzzy numbers by using the given ranking method based on generalized intuitionistic fuzzy costs. This new algorithm is very easy to understand and it can be used for all types of assignment problems (Gani, & Mohamed, 2015). Operations Research has been used to build the assignment models and design algorithms of optimization by Fan Wang et al. Model deals with the optimization aspect of the paper assignment problem. Earlier researchers work was based on global properties of the assignment whereas the researcher's algorithms are based on min-cost matching/flow, integer programming, or heuristics without provable guarantees (Wang et al., 2013). Instead of considering the fairness of the assignment as an objective; the fairness is treated as the Leximin criterion for the problem which are NP-hard without provable guarantees (Wang et al., 2013). In this technique fairness is captured by the Leximin criterion instead considering the fairness of the assignment as an objective. For the variants of the problem that are NP-hard researcher claims to be resorted to approximation algorithms with worst-case guarantees (Wang et al., 2013).

D. Yimam et al. proposed two-stage language model to expert search. It consists of two parts, relevance model and co-occurrence model. Each of these sub-models is based on extraction of one type of metadata. Relevance of query document is modeled through developed query model. The co-occurrence model represents whether or not the query is associated with a person and both models are based on statistical language modeling (Yimam & Kobsa, 2005). Experiments were conducted for testing the effectiveness of model and it is observed that model performs better with the use of two-stage model than the existing techniques that solely use co-occurrence between keyword and expert (Yimam & Kobsa, 2005). It is also suggested that the window-based sub-model should be preferred over the document-based sub-model. Study reveals that among all possible combinations, the best combination achieving maximum performance is sub-models based on both body-body and title-author co-occurrence. The performance measured in terms of average precision is improved in case of clustering-based re-ranking. Biswas et al. have used data mining to extract the keyword-list by doing unsupervised clustering or supervised learning by using pervious accepted papers as training set. Researchers have compared the applicability of variety of content-based filtering and suggested that hybrid approaches is a better comprehensive way as compared to others (Biswas & Hasan, 2007). Researchers used the powerful Vector Space Modeling (VSM) technique and machine learning tools extracting features such as key phrases, and Semantic Web technology such as ontology-driven topic inference to facilitate an efficient assignment of reviewer to papers. For computing relevance of assignment of paper to reviewer with the help of VSM, researchers have made use of ADL conference having set of 10 papers, 30 reviewers and paper reviewer assignment information. Researchers claim that combining domain knowledge with automatically extracted keywords that is ontology-driven topic inference using could potentially identify the most relevant reviewers for a paper (Biswas & Hasan, 2007).

Researcher Watanabe et al. has build a scale-free network with keywords as vertices representing reviewers' expertise and the similarity between manuscripts' topic, and two keywords as the probability of connecting between the respective vertices. The matching degree is the weighted average of similarities between each pair of keywords of manuscripts and reviewers (Watanabe et al., 2005). Researchers Rodriguez and Bollen had built a co-authorship network with vertices representing experts, edges representing cooperativeness between two experts, and weights representing the strength of relation Instead of keywords (Rodriguez et al., 2007).

S. Researchers Hettich et al. developed a prototype application to assign experts for reviewing the funding proposals. They have modeled the reviewer assignment as a retrieval problem and to calculate the match score between reviewer candidates and proposals, the TF-IDF weighting is used (Hettich & Pazzani). Wenbin et al. has modeled the reviewer assignment problem as constrained optimization framework. The constraints are to be formalized as penalty in the objective function. We can also use these constraints for optimization solving process. Researchers claim that, for solving the optimization framework transforming the problem to a convex cost network flow problem guarantees an optimal solution (Tang et al., 2012). Experimental results tested with two datasets demonstrated that the proposed approach can effectively and efficiently match experts with the queries and feedbacks from the users were observed to be very positive.

An optimization approach based on preference matrix and the model of Asymmetric Travelling Salesman Problem (ATSP) is proposed by researchers Yanqing Wang et al. Researchers have defined the assignment between levels of reviewers to a level of authors as a distance as they believe that different learning outcomes will be achieved when we define the assignments between the students with different competence levels (Wang et al., 2014).

Researchers further claim that the maximum learning outcome is obtained when the total distance is minimized, (Wang et al., 2014). Researchers claim that the practical value of the optimization approach and the performance of the assignment with preference matrix is better than that of the random assignment. Further authors suggested scope for further research as – rule matrix can be designed, the orientation effect of different rule matrixes on the competence improvement of organizations. In general it is noticed that existing methods for expertise matching mainly fall into two categories: probabilistic model and optimization model. C. B. Haym et al. have developed probabilistic model. This model improves the matching accuracy between experts and queries. Model is based on keyword matching, latent semantic indexing, and probabilistic topic modeling (Haym et al., 1999). Researchers M. Karimzadehga et al. have treated submissions as a combination of multiple subtopics. It is noticed that all constraints are not fulfilled and constraints by heuristics are only considered. The optimization model tries to incorporate the constraints as a component in an optimization framework such as integer linear programming and minimum cost flow (Karimzadehgan et al., 2008) (Karimzadehgan & Zhai, 2009).

Artificial intelligence algorithms are analyzed by T. Kolasal et al. for paper reviewer assignments problem. They studied Genetic Algorithms (GA), Ant Colony Optimization (ACO), and TABU search (TS), and the performance is tested (Kolasa & Krol, 2011). They proposed two hybrid methods: the ACO-GA and GA TS algorithms, and conducted computational algorithms using different data sets. Researcher claim that the hybrid methods when combined efficiently; they are effective and achieve good results (Kolasa & Krol, 2011).

Researchers Sixing Yan et al. have proposed an approach of paper-reviewer assignment considering potential Conflict of Interests (CoI) between authors of submitted papers and reviewers. Initially, academic networks of scholar-to-scholar and institution-to-institution are extracted using academic activities records. Then, conflict of interests is computed using paths distance between related authors and reviewers from academic social networks. Topic relevance is computed using similarity between submitted paper and reviewer's publications. Finally, the paper-reviewer assignment approach is modeled as a minimum cost maximum flow problem to maximize topic relevance and to minimize Conflict of Interests (Yan et al., 2017). When tested with a huge real scholar data and live conference the approach was evaluated and was found effective. Authors Li et al., have implemented a new review assignment system, named as 'Erie'. Erie is based on best practices in the literature and own improvements and experimental evaluations. Latent Semantic Indexing is used to compute manuscripts similarity with each of the reviewer's sample published papers. The maximum score is used for assignment of reviewer to paper (Li & Hou, 2016). A constrained optimization problem based on matrix of suitability scores, is solved so that the total suitability of all papers is maximized satisfying workload and CoI constraints. Erie has been successfully used for two thirds of the reviews assigned in INFOCOM 2015, and all of the reviews assigned in INFOCOM 2016. It is observed convincing anecdotal and statistical evidence that Erie has noticeable improvement in the quality of reviews in INFOCOM's new double-blind review process.

Most of the researchers have traditionally used different factors to choose appropriate reviewers. These factors include – relevance between reviewer candidates and submissions and the diversity of candidates. However, often the facts are ignored that – there could be changes of reviewer interest and tread of reviewers' interest may change over the time (Jin et al., 2017). X Li et al. claimed that recent publications are the most authenticate and strong source for representing and extracting interest/expertise of experts. They have considered a time interval and the recent publications are assigned the higher weight than older publications. Expertise labels and reviewers' recent publications are exploited well in this approach (Li & Watanabe, 2013). However, it is difficult to obtain in increasingly narrow research fields to extract the review interest. The selected keywords/domain by reviewers are modeled as an interest trend and direction and the smoothness are recognized.

Jian Jin et al. considered three important facts- relevance between reviewer candidates and submissions, the interest trend and the authority of candidate. An integer linear programming problem is formulated with extracted aspects. Researchers have tested the performance with two large datasets that are built from WANFANG and ArnetMiner. Results reveal the availability of the proposed approach in modeling the temporal changes of reviewers' research interest demonstrating the effectiveness of approach (Jin et al., 2017) (Cao et al., 2005).

The study has noticed few problems related to the peer review process and authority of reviewers. Some of these are not yet addressed in present reported research. It is further noticed that few of the issues are addressed but are handled independently. It is observed that there is correlation and dependency among few of the issues. Hence, cause-effect and interrelationship needs to be addressed well (Jan & Contreras, 2015). List of such identified issues and challenges related to RAP are elaborated and listed in forthcoming summary section of this paper.

11.4 Performance Evaluation Techniques and Data Sets Used For Testing

Typically precision and recall are used as performance measures in information retrieval systems. But these are not found suitable by most of researchers. The study reveals that researchers have proposed different evaluation metrics according to the problem definition of their own.

Table 11.2 Performance Measures popularly in use

Matching Score	Average Percentage of Key Word Matches Between the Expertise and Papers
Load	Maximum number of papers to be reviewed by reviewer
Load Balance	No reviewer is overburdened with papers
Load variance	Variance of the number of papers assigned to different reviewers
Expertise	Measure of reviewers specialty in the specific domain
Relevance	Evaluates the topical similarity between a reviewer and a submission
Interest Trend	The interest trend of a reviewer, which evaluates the degree of candidates' willingness to review a submission
Authority	Measure of reviewers recognition in the larger scientific community
Diversity	Measure of reviewers diverse research interests and background
Confidence	Level of confidence intimated by reviewer and is integrated as parameter in computation.
Coverage	Every paper should be reviewed a sufficient number of times.
Fairness	Treating equally without bias or favoritisms or discrimination.
Quality	To be measured in terms of F-score / accuracy
RMSE	Root Mean Square Error Value by checking the relevant Assignments based on COI
Precision	Precision is a description of random errors / a measure of statistical variability.
Accuracy	Accuracy is the closeness to the true value. Accuracy consists of Trueness and precision (proximity of measurement results to the true value)
F-score	Measure of a test's accuracy. It considers both the precision and the recall of the test to compute the score. Also defined as the weighted harmonic mean of Diversity and Coverage measures.
Unanimity	An efficiency of a peer-review process can be defined in terms of unanimity. Unanimity states that when there is a common agreement among all reviewers, then the aggregation of their opinions must also respect this agreement. Two measures – Pair-wise Unanimity and Group Unanimity (PU & GU respectively)

Table 11.3   Data Sets used for performance testing

TIManG-Text Information Management and Analysis Group has created data set for evaluating multi-aspect review assignment	http://sifaka.cs.uiuc.edu/ir/data/review.html
CIKM	www.cikmconference.org
"Enterprise Track" in TREC	https://trec.nist.gov/
ACM SIGIR (proceedings since 1971 from the ACM digital library)	https://dl.acm.org/
NIPS (N10 and N09, from the 2010 and 2009 editions, respectively, of the NIPS conference, one of the leading conferences in machine learning)	http://nips.cc
WANFANG	https://www.jiscmail.ac.uk/cgi-bin/webadmin?A0=DATA-PUBLICATION
ArnetMiner	http://review.arnetminer.org
NTU	http://www.ntu.edu.sg/home/axsun/datasets.html
ICLR 2017 Submissions	https://openreview.net/submissions?id=ICLR.cc/2017

Performance evaluation measurement of certain parameters helps to know how effective it is and is based on performance metric and sample datasets. Different metrics may be used depending upon model and dataset. Choice of metrics plays an important role in measuring and comparing the performance of machine learning algorithms. Popularly for an unbiased evaluation, dataset is split into three parts: a training, test, and validation set. Data sets used by most of the researchers are listed below in Table 11.3

11.5 Brief Summary

Satisfactory number of research papers is devoted for the domain-Reviewer-assignment Problem. Study of literature reveals that popularly the RAP process is divided broadly in two phases: Reviewer-Paper Matching and Reviewer-Paper Assignment.

The first phase is to compute and rank the match among group of reviewers and submitted papers. This is popularly done either using information retrieval techniques or by having the reviewers expressing their preferences.

It is often handled by making reviewers and paper authors select a set of keywords that best describe their interest and expertise. Often to fulfill the conflict of interest constraint, either self declaration or DBLP website is mined for bibliographic information of experts. For each reviewer information extracted is generally – publication titles and co-authors, bibliographic information, reviewer's thesis and thesis supervised to locate research scholar and his/her supervisor.

The second phase in the paper assignment problem is the optimization process itself, assigning paper to referees so as to maximize match, optimal assignment satisfying maximum constrains.

It is revealed from literature study that most research mainly focus on the relevance between paper and reviewer in reviewer assignment or matching with some practice constraints. Most of them are borrowed from Information Retrieval algorithms. Major parameters considered for assignment include – topic relevance, topic coverage for submissions, research impact of publications, diverse background of experts, semantic features of reviewer candidates, the relations between candidates and authors, research interest trends of reviewer, reviewers' willingness on submissions, preferences of reviewer, Conflict of interest, Maximum number of publications willing to review, time deadlines, and timely completions. However, the complex relationships between authors of paper and reviewer, like conflict of interest, unavailability of updated profiles of reviewers, unawareness among experts, level of confidence, payment and rewards, gender, anonymity, reviewers subjectivity, reviewers suggested by authors and authenticity, Internationalization are seldom concerned are yet to be addressed well by researchers. The system must satisfy very good balance between goodness and fairness demands of reviewer-paper assignment process.

Popular Methods used for first phase include – Firstly, computation of reviewer-paper match. Few approaches use Latent Semantic Indexing by making reviewers and paper authors to select several keywords from the same pre-established list that tries to cover all subjects in the conference domain. Often the reviewer profile is extracted from their online published papers, and it is then matched using LSI and papers are assigned based on matches. Major difficulty associated with these techniques is that most of the reviewers do not select keywords or do not appropriately select keywords. Hence the algorithm cannot guarantee that there is at least one keyword matching between the reviewer and the paper. This often leads to problem of receiving irrelevant papers for review at reviewer end leading to disappointed. Few authors have used vector space modeling (VSM) technique and machine learning tools extracting features such as key phrases, and Semantic Web technology. But again scarcity and incomplete information leads to poor assignments. Commonly used variables to determine a reviewer's expertise are – PhD Thesis Year, PhD Supervised, Books, Book Chapters, Activity, Research Areas, Publications, Citations, H-index, Sociability, Diversity, and Professional Age.

Popular Methods used for phase 2 use Probabilistic, Heuristics, Minimum-cost-flow model, linear programming and Optimization approaches. Optimization approach formulates RAP as a scheduling problem, where resources are the paper themselves and the "jobs" are the reviewers. Few researchers exploited Operations Research based approaches to build the assignment models and design algorithms of optimization. Some of the techniques are based on Machine learning algorithms, Fuzzy, Greedy, Evolutionary approach and few have also used hybrid approach.

11.5.1 Research Opportunities – Gap and Challenges

Study reveals that there are various constraints to be satisfied

- Avoiding conflicts of interest between the reviewer and authors,
- Paper should be reviewed by person with expertise/interest in paper domain,
- Each reviewer load cannot exceed a certain number of papers, and
- Each paper should be reviewed a sufficient number of times.

Study reveals that there is still enough scope for improvement and yet many challenges are to be handled, of which few are listed below

- There are no clear norms for reviewer-paper assignments,
- There are no full proof system that fairly matches the paper and reviewer.
- There is variation in parameters and its preference in assignment process.
- There is no way to authenticate the information extracted from reviewer's profiles and publications.
- There is variation in payments and rewards against reviews.
- There is unawareness among experts about call for reviewers.
- There is unwillingness among experts towards review process.
- Often reviewers do not complete the reviews.
- Often reviewers do not complete reviews in time.
- Huge number of papers is received near deadlines.
- There is scarcity of good reviewers.

- There are issues like little compensation, recognition, internationality, gender and seriousness of reviewers that affect whole process.
- There is problem of anonymity – there are experts in domain but are not known to the world may be because they do not publish or do not popularize themselves.
- Often reviewers do not fill in complete information and do not select appropriate keywords or domains of interest.
- Recent trends and feedback of reviewers is not effectively used for assignments.
- There are no benchmarks for performance assessment of reviewer-paper assignment systems. Existing techniques are either subjective or objective; none of them clearly measure performance.
- Proper training or guidelines is not provided to reviewers.
- There is need of guidelines to handle improper assignments. Improper assignments lead to unfair reviews, reassignment and slower down the whole process.
- There is often dissatisfaction among authors due to unclear reviewer comments or justification, sometimes weak papers are published and relatively better papers are rejected.
- Feedbacks of authors and reviewers are to be used for improvement in process and system.

Major gaps and challenges are listed here. With faster technology upgrades additional challenges may evolve that are to be addressed by researchers.

11.6 5G and Reviewer Assignment Problem

With the paradigm shift to 5G, there will be noticeable application implications to the adoption of 5G in conference management systems and related activities. Looking at growth in number of conferences scheduled, number of papers submitted, use of technology, popularity of web based systems with interface of mobile devices, need of distant effective communication, urge of timely completions there is necessitate of proficient management of the resources including web resources. For efficient conference management, there is demand of high data rate, affordable cost, low energy, system with high capacity, low latency, and most importantly the connectivity among massive devices. Adaptation of 5G can serve the purpose. In developing countries, there is need to create awareness for adaptation of potential use of 5G, Artificial Intelligence (AI), Block Chain Technologies (BCT) and ICT ecosystem.

11.7 Conclusions

The study indicates that there is significant contribution of researchers in developing efficient solutions to reviewer assignment problem. Paper summarizes many challenges in domain and potential open areas for future research. The contributed solutions have many limitations indicating that there is enough scope to improve the performance and to standardize the process. Adaptation of 5G with AI, BCT and use of machine learning algorithms can help is designing proactive algorithms to solve reviewer-assignment problem fulfilling the future needs and in addressing the challenges listed in this paper.

References

Biswas, H. K. and Hasan, M. M. (2007) Using Publications and Domain Knowledge to Build Research Profiles: an Application in Automatic Reviewer Assignment. *International Conference on Information and Communication Technology*, pp. 82–86.

Cao, Y., Liu, J. Bao, S., and Li, H. (2005). Research on expert search at enterprise track of TREC 2005, TREC 2005, Voorhees and Buckland, p. 4.

Dumais, S. and Nielsen, J. (1992). Automating the assignment of submitted manuscripts to reviewers, *Research and Development in Information Retrieval*, pp. 233–244.

Dumais, S. T. and Nielsen, J. (1992). Automating the assignment of submitted manuscripts to reviewers, *Proceedings of the 15th annual international ACM SIGIR* conference on Research and development in information retrieval, ACM, New York, NY, USA, pp. 233–244.

Gani, A. N. and Mohamed, V. N. (2015). An algorithm for solving assignment problems with costs as generalized trapezoidal intuitionistic fuzzy numbers. *Inter. J. of Pure and Appl. Math*, 104(2), 561–575.

Hartvigsen, D., Wei, J. C., and Czuchlewski, R. (1999). The conference paper-reviewer assignment problem. *Decision Sciences*, 30(3), pp. 865–876.

Haym, C. B., Hirsh, H., Cohen, W. W., and Nevill-manning, C. (1999). Recommending papers by mining the web, *Proceedings of the 20th international joint conference on Artificial intelligence*, pp. 1–11.

Hettich, S. and Pazzani M. J. (2006). Mining for proposal reviewers: Lessons learned at the national science foundation, *KDD '06*, pp. 862–871.

Jan, A. U. and Contreras, V. A. (2015). A List of Research Problems Encountered in the Peer Review Process. *Saber y Hacer*, 2(1), 114–125.

Jin, J., Geng, Q., Zhao, Q., and Zhang, L. (2017). Integrating the Trend of Research Interest for Reviewer Assignment, International World Wide Web Conference Committee, (IW3C2), published under Creative Commons. pp. 1233–1241.

Karimzadehgan, M. and Zhai, C. (2009). Constrained multi-aspect expertise matching for committee review assignment, *Proceedings of the 17th ACM international conference on Information and knowledge management (CIKM'09)*, pp. 1697–1700.

Karimzadehgan, M., Zhai, C., and Belford, G. (2008). Multi-aspect expertise matching for review assignment, *Proceedings of the 17th ACM international conference on Information and knowledge management (CIKM'08)*, pp. 1113–1122.

Kolasa, T. and Krol, D. (2011). A Survey of Algorithms for Paper-reviewer Assignment Problem. *Taylor and Francis – IETE Technical Review*, 28(2), pp. 123–134.

Li, B. and Hou, Y. T. (2016). The New Automated IEEE INFOCOM Review Assignment System. *IEEE Network,* 30(5), 18–24.

Li, X. and Watanabe, T. (2013). Automatic paper-to-reviewer assignment, based on the matching degree of the reviewers. *Procedia Computer Science*, 22, 633–642.

Mehlhorn, K., Garg, N., and Kumar, A. (2010). Assigning Papers to Referees – Objectives, Algorithms, Open Problems, *Algorithmica*, 58(1), pp. 119–136.

Merelo-Guervós, J.J. and Castillo-Valdivieso, P. (2004). Conference paper assignment using a combined greedy/evolutionary algorithm, Parallel Problem Solving from Nature—PPSN VIII, 3242, pp. 602–11. Springer.

Mimno, D. and McCallum, A. (2007). Expertise modeling for matching papers with reviewers. *Proceedings of the 13th ACM SIGKDD International Conference on Knowledge Discovery and Data Mining (SIGKDD'07)*, pp. 500–509.

Rodriguez, M. A., Bollen, J., and Van de Sompel, H. (2007). Mapping the Bid Behaviour of Conference Referees, *Journal of Informetrics,* 1(1), pp. 62–82.

Shah, N. B., Tabibian, B. Muandet, K., Guyon, I. and Luxburg, U.V. (2017). Design and Analysis of the NIPS 2016 Review Process, *Computing Research Repository*, pp. 25.

Tang, W., Tang, J., Lei, T., Tan, C., Gao, B., and Li, T. (2012). On Optimization of Expertise Matching with Various Constraints, *Elsevier International Journal of Neuro computing*, 76(1), pp 71–83.

Wang, F., Zhou, S., and Shi, N. (2013). Group-to-group reviewer assignment problem. *Computers and Operations Research*, 40(5), 1351–1362.

Wang, Y., Jiang, Y., Wang, X., Zhang, S., and Liang, Y. (2014). Solving reviewer assignment problem in software peer review: An approach based on preference matrix and asymmetric TSP model, *Computing Research Repository*, pp. 11.

Watanabe, S., Ito, T., Ozono, T., and Shintani, T. (2005). A Paper Recommendation Mechanism for the Research Support System Papit, *International Workshop on Data Engineering Issues in E-Commerce*, pp. 71–80.

Yan, S., Jin, J., Geng, Q., Zhao, Y., and Huang, X. (2017). Utilizing Academic-Network-Based Conflict of Interests for Paper Reviewer Assignment, *International Journal of Knowledge Engineering*, 3(2), pp. 65–73.

Yarowsky, D. and Florian, R. (1999). Taking the load off the conference chairs: towards a digital paper-routing assistant. *Proceedings of the 1999 Joint SIGDAT Conference on Empirical Methods in NLP and Very-Large Corpora*, pp. 220–230.

Yimam, D. and Kobsa, A. (2005). Demoir: A hybrid architecture for expertise modeling and recommender systems. *Engineering Issues in E-Commerce*, pp. 71–80.

Index

A

Advanced Encryption Standard (AES) 248
Ant Clustering Algorithm 218
Active Ethernet 96, 97, 101
Advanced Manufacturing 56, 57, 59
Argentine Data Protection Regulations 20
ARPU 12
Artificial Intelligence 1, 28, 49, 70
Asymmetric Travelling Salesman Problem 292
Augmented Reality 1, 4, 9
Autonomous Driving 1

B

Balloting 121, 129, 130
Biometric Register 127
Biometric Verification Devices 121, 128, 136, 138
Big data 5, 28, 70, 74
Big Data Analytics 171, 175, 178, 180
Blackhole attack 228, 229
Blockchain 1, 117, 123, 136, 151
Blockchain Technology 10, 123, 144
Brazilian Civil Code 21

Brazilian computer industry 5, 38, 42, 47
Brazilian Industry Agenda 4.0 56
Brazil's National Computer Policy 37, 38, 39, 42
Broadband 1, 20, 30, 85
Broadband access 85, 96, 106, 111
Broadband connectivity 86, 95, 97, 102
Broadband services 3, 27, 86, 92

C

CAGR 51, 52, 53, 54
Capital Expenditure (CAPEX) 96, 153
CCTV 244, 245, 246
Cloud Architecture 15, 147, 155, 168
Cloud adoption 31, 147, 149, 154
Cloud Computing 4, 15, 20, 27
Cloud Based Services 4, 5, 6
Cloud Quality of Service Metrics 168
Cloud Service Providers 16, 19, 20, 32
Competition 12, 16, 41, 48
Computer Policy 5, 37, 39, 42
Computer Supported Cooperative work 53

Computational
    Intelligence 217
Commercial IOT 240
Community Cloud 149, 151, 152
Consensus 40, 123, 125, 131
Consumer IOT 240
Cooperatives 5, 85, 88, 91
Cooperative intrusion 219
Cooperative Jamming (CJ) 251
Cyberattacks 33
Cybercrime 18, 19, 30, 203
cyber-physical systems 49
Cyber security 3, 7, 216
Cryptocurrency 124, 125

D

Dark fiber 92
Datacenter 148, 150, 154, 159
Data-driven economy 67
Data integration 7, 171, 175, 179
Data Management 15, 26,
    173, 270
Data Privacy laws 17
Data processors 25, 49, 70
Data processing 26, 43, 58, 63
Data Protection
    Regulation 5, 20, 24, 63
Data Protection
    Supervisor 65, 78
Denial of Service 3, 204,
    215, 219
De-regulated markets 93
Developing countries 1, 41,
    58, 63
Differential Power Analysis
    (DPA) 75, 260
Digitalization 11, 15, 26, 31
Digital Agenda 4, 11, 12, 15

Digital natives 69
Direct Recording
    Electronics 121, 122
District Health Information
    Management System
        (DHIMS II) 269, 271,
            275, 277
Digital literacy 139
Digital strategy 24
Digital Single Market 68, 78
Digital transactions 123
Digital Voters Register 133,
    134, 136
Distributed Ledger 123, 124, 141
Domain Name System
    (DNS) 161

E

E- Government Services 15, 145
E-Health 26, 242, 243
E voting 118, 120, 122, 133
Economic growth 31, 39, 42, 64
Effects of spillover 55
Electronic Medical Records 159
Electric cooperatives 87, 89,
    97, 99
Electric co-ops 92
Electoral process 117, 129,
    131, 141
Elections 6, 66, 117, 120
Electronic Voting 117, 118,
    119, 121
Electronic Voting Machine 118,
    121, 141
Elliptic Curve Cryptography
    (ECC) 248
Energy harvesting 248
European Commission 64, 73, 78

F

Facebook 26, 64, 69, 73
Federal Broadband
    Programmes 111
Federated Blockchain 125,
    131, 132
Fifth Generation (5G) 1, 11,
    27, 110
Finality 125, 127
File Transfer Protocol 222
Fourth Generation (4G) 1, 2, 12
Free trade 20, 30
FTTC 112
FTTH 97, 112, 114
Fully Allocated cost 102
Full Duplex 248, 252, 265

G

Google's advertising platform 70
General Data Protection
    Regulation (GDPR) 5, 22,
        63, 78
Generation Z 69
Gigabit Passive Optical Networks
    (GPON) 90, 91, 98, 101

H

Health Data
    Management 71, 269,
        272, 277
Health Information
    System 269, 270, 272, 274
Health Information
    Technology 269, 272,
        275, 284
Host-based IDS 210
Host-based IPS 212

Human Resource and
    Accounting Information
        Systems 159
Human-to-machine 1
Hybrid Fiber-Coaxial 97
Hybrid Cloud 147, 148,
    151, 154
Hybrid Cloud
    Architecture 147, 148,
        155, 168

I

ICT 1, 5, 242, 299
Immutability 125, 127
Independent telecommunication
    providers 93
Industry 4.0 2, 37, 49, 54
Industrial development 38, 39,
    40, 50
Industrial IOT 240, 249, 264
Industrial Policy 37, 39, 42, 48
Information Systems 7, 29,
    153, 161
Infrastructure as a Service 151
Innovation 7, 15, 29, 40
Internet of Things (IoT) 1, 12,
    28, 37
Intrusion detection System 202,
    207, 214, 226
Intrusion prevention system 203,
    210, 214, 224
IS Success Model 7, 269,
    272, 276

J

JSON 174

K

Keynesian 40, 57
Knowledge management
generation 139, 175, 177, 301

L

Local Area Network (LAN) 160,
162, 212, 222
Ledger Technology 123

M

Machine Learning 1, 8, 70, 115
Machine-to-Machine
communications 1, 4
MapReduce functionality 195
Marginal cost 150
Miners 124
Minimum Viable
Architecture 156
Mobile subscribers 70
Municipal networks 86, 95
Multiple-Input-Multiple-
Output (MIMO) 252

N

National Broadband Plan 20
National Development Plan 42
National Industrial Policy 4.0 37
National IoT Plan 55
Net Present value 95
Network Security 201,
207, 230
Network-Based IDS 121, 207,
210, 217
Network-Based IPS 203,
207, 212, 225
Neo-liberalism 47
Next Generation
Technology 86, 92

NGT provider 6, 85, 86, 94
Non-Governmental
Organization 95
Non-Profit
Cooperatives 86, 94

O

Open Ledger 124
Optical Distribution Network
(OPN) 101
Operational Expenditure
(OPEX) 96, 102, 153
Optical Line Terminal
(OLT) 97, 100, 101
Optical Network Unit
(ONU) 97

P

Peer-to-Peer network 123
Personal data 5, 18, 21, 23
Physical Layer attack 237,
247, 253
Platform 4, 8, 20, 37
Platform as a Service 150, 151
Privacy 3, 18, 21, 24
Privacy paradox 68, 84
Private Blockchain 125, 131, 137
Private Cloud 151, 154,
161, 166
Provenance 124, 125,
126, 182
Public Blockchain 125
Public Cloud 16, 148, 151, 154
Public Key Infrastructure 165

Q

Quality of Experience 2, 96
Quality of Service 2, 6, 8, 96

R

Random Access Memory 151, 159, 162

Redundant Array of Inexpensive Disks (RAID) 157, 222

Reinforcement learning 205, 206

Regulation 5, 9, 12, 16

Reviewer-Paper Assignment 288, 289, 296, 298

RFID 249, 252, 255, 258

Rural broadband 86, 88, 92, 96

Rural broadband services 86, 87, 88, 92

Rural Electric Networks 99

Rural NGT providers 85

S

Search Engine Optimization 173, 199

Self-Supported-All-Dielectric 99

Semi-supervised learning 205, 206

Service Level Agreements 30, 153, 154

Software as a Service 26

Schumpeterian 40

Side channel attack 251, 260, 261

Silicon Valley 40, 71, 198

Simple Power Analysis (SPA) 206

Small and medium-sized enterprises (SME) 3, 63, 73, 175

Smart city 242

Smart devices 237, 238

Smart grid 6, 88, 93, 98

Smart grid network 93

Smart village 242

Social Media 7, 69, 74, 171

Social Media Metrics 7, 171, 174, 177

Spoofing 207, 215, 225, 249

Sybil attack 203, 221, 225, 227

System Quality 269, 273, 277, 281

Synthetic biology 49

Sub Saharan Africa 147, 148

T

Telecommunications networks 85

Threat analysis 237

Throughput 157, 158, 161, 163

Trust 3, 6, 13, 68

Trust by default 117, 118, 122, 125

TV Distribution Networks 93

U

Universal Declaration of Human Rights 12

Utility cooperatives 5, 85, 87, 91

V

Vector Space Modeling 292, 297

Vertical Industrial 47, 54

Virtualization 149, 150, 151, 156

Virtualized hardware 150

Virtual Machine 150, 159, 162, 165

Virtual Personal Network (VPN) 161

VoIP 159

W

Web servers 222
Wireless Area Network (WAN)
    160, 162, 163, 167
Wi-Fi 29, 160, 240
Wireless Sensor Network 201,
    203, 207, 215

World Wide Web 181, 211, 301
Wormhole attack 225, 226,
    227, 228

X

XML 134, 136, 174

About the Authors

Aboli H. Patil is a Research scholar at Smt. Kashibai Navale College of Engineering, Pune. Her area of research is Machine Learning and Data Mining. She is perusing PhD under the guidance of Dr. Parikshit Mahalle. She has total 06 years of teaching experience. She is co-author of two books Fundamentals of Programming Language and Theory of Computation. She has 1 copyright and 1 patent (Filed) to her credit. She is member of various professional bodies.

Bright Kodua is a public health officer (Informatics) and IT consultant at the Holy family Hospital. He holds a Bachelor in Public Health Informatics from the Catholic university in Ghana. He is incharge of Records management, Data Validation, Disease Surveillance through data, Performance Reviews, Operational Research, System Quality Assurance at the hospital.

Prof. Darío M. Goussal is a full-time professor, founding member and coordinator of the Rural Telecommunications Program in the Department of Electricity and Electronics, School of Engineering of Universidad Nacional del Nordeste (UNNE) at Resistencia, Argentina, where he has been teaching since 1980. He has conducted many research projects about rural community telecenters, expansion behavior of local networks in small towns, supply and demand studies in optical backbone routes, National Research and Educational Networks (NRENs), white-space technologies with cognitive radios, waste electric and electronic equipment (WEEE) and feasibility aspects of rural wireless and optical broadband networks operated by community providers and utility cooperatives, having authored about a hundred of scientific and technical publications. A specialist in problems of the strategic planning of Rural Telecommunications and Senior Expert of the International Telecommunications Union (ITU), he has participated in engineering, training, consulting, project evaluation and research assessment missions for national government agencies, regional international organizations or agencies in over 20 countries.

Dr. Ezer Osei Yeboah-Boateng a Telecoms Engineer and ICT Specialist with over twenty-five (25) years' experience in telecommunications switching systems, cyber-security, digital forensics, business development, digital Transformation, project management, change management, knowledge management, strategic IT-enabled business value creation and capabilities to develop market-oriented strategies aimed at promoting growth and market share. His research interests are in cyber-security, Big Data, Cloud Computing, Social Media and Fuzzy Systems, etc. He is currently the Dean of the Faculty of Computing & Information Systems, at the Ghana Technology University College (GTUC). He serves on the governing boards of the National Information Technology Agency (NITA) in Ghana, and the Ghana Technology University College (GTUC).

Dr. Geetanjali R. Shinde Graduated in Computer Engineering from Pune University, Maharashtra, India in 2006 and received Master's degree in computer Network from Pune University in 2012. In 2018 awarded PhD in Wireless Communication, Department of Electronic Systems, Center for Communication, Media and Information Technologies Copenhagen, Aalborg University Denmark. Since January 2008, she is working as an Assistant Professor in Department of Computer Engineering, STES's Smt. Kashibai Navale College of Engineering, Pune, India. She has published 25+ papers at National and International level. She received research funding from Board of College and University Development, SPPU, Pune, India.

Dr. Idongesit Williams is a Post-Doctoral Research fellow at Aalborg University. He holds a Ph.D in Information and Communication Technologies, specializing in telecommunication and Internet infrastructure development and policies. He has for 7 years worked as a Lecturer at Aalborg University, Copenhagen where he taught Internet Governance and Economics. Currently, he is involved in EU projects. His current project is on the promotion of digitization in the Baltic Sea region. In addition to his core area of specialization, he has conducted research into other areas involving the diffusion and adoption of ICTs in organizations and by Individuals. He has published widely with more than 40 publications on research areas ranging from Public Private Partnerships in the development of Broadband infrastructure, ICT business modeling, Knowledge management and organizational learning, Gender, ICT, digitalization in the EU and entrepreneurship among others. He has served as guest editors for Special issues in Journals. He is also a consultant for SMEs on the development of requirement specifications and use cases for IT based services.

Kenneth Kwame Azumah is Ph.D. fellow at CMI (Center for Communication, Media and Information Technologies) at Aalborg University, Copenhagen, since autumn 2016. He attended the Kwame Nkrumah University of Science and Technology, Ghana where he received his B.Sc. in Computer Science in 2001. Azumah received an M.Eng. in Electrical Engineering and Information Technology from Deggendorf Institute of Technology, Germany in 2009 and an MBA from the Blekinge Institute of Technology, Sweden in 2011. He is currently working to complete his Ph.D. at Aalborg University, Copenhagen where his research centers on hybrid cloud computing with process mining.

Mahendra B. Salunke is a graduate in Electronics from North Maharashtra University, India in 1998 and Post Graduation in Computer Science and Engineering from Visvesvaraya Technological University, India in 2008. He has registered for a Doctorate of Philosophy (PhD) in Computer Engineering at Shrimati Kashibai Navale College of Engineering, Pune, Research Center affiliated to Savitribai Phule Pune University, India. His field of specialization is "Embedded Systems and Embedded Security". He is performing his research work in providing lightweight embedded security for resource-constrained devices used in machine-to-machine communication. He is having three copyrights on his name and applied for one patent in India. He has published more that 20 technical papers at national and international level. He is a Life Member of Computer Society of India, Indian Society for Technical Education and Member of Association of Computing Machinery (ACM). He is having more than 20 years of experience in industry and academia. Currently, he is working as Assistant Professor in Department of Computer Engineering, Pimpri Chinchwad College of Engineering and Research, Ravet, Pune.

Marcelo Castellano Lopes, holds a B.A. in Economics. M.A. candidate in the Graduate Program in Public Policies at Federal University of Paraná, Brazil.

Prof. Marcelo Vargas is a Professor in the University of Paraná State, Brazil. He obtained an M.A. in Economics from Graduate Program in Economic Development at Federal University of Paraná. Ph.D. candidate in the Graduate Program in Public Policies at Federal University of Paraná.

Dr. Pankaj R. Chandre obtained his B.E degree in Information Technology from Sant Gadge Baba Amravati University, Amravati, India and M.E. degree in Computer Engineering from Mumba1 University Maharashtra, India in the year 2011. Currently he is persuing his PhD in Computer Engineering from Savitribai Phule Pune University, Pune, India. He is currently working as an Assistant Professor in Department of Computer Engineering, STES's Smt. Kashibai Navale College of Engineering, Pune, India. He has published 60 plus papers at international journals and conferences. He has guided more than 30 plus under-graduate students and 20 plus postgraduate students for projects. His research interests are Network Security & Information Security.

Dr. Parikshit N. Mahalle, obtained his B.E. degree in Computer Science and Engineering from Sant Gadge Baba Amravati University, Amravati, India and M.E. degree in Computer Engineering from Savitribai Phule Pune University, Pune, India. He completed his Ph. D in Computer Science and Engineering specialization in Wireless Communication from Aalborg University, Aalborg, Denmark. He has more than 18 years of teaching and research experience. He has been a member board of studies in computer engineering, Savitribai Phule Pune University (SPPU), Pune, India. He has been a member – Board of studies in computer engineering, SPPU. He is member – BoS coordination committee in computer engineering, SPPU. He is also serving as technical program committee for International conferences and symposia. He is IEEE member, ACM member, Life member CSI and Life member ISTE. He has 6 patents. He has authored 7 books on Identity Management for Internet of Things, (River Publishers), Identity Management Framework for Internet of Things, (Aalborg University Press), Data Structures and Algorithms, (Ceengage Publications), Theory of Computations, (Gigatech Publications), Fundamentals of Programming Languages – I, (Gigatech Publications) and Fundamentals of Programming Languages – II, (Gigatech Publications). Design and Analysis of Algorithms: A Problem Solving Approach, (In Process) – Cambridge University Press.

He is also the recipient of "Best Faculty Award" by STES and Cognizant Technologies Solutions. He has also delivered invited talk on "Identity Management in IoT" to Symantec Research Lab, Mountain View, California.

Currently he is working as Professor and Head in Department of Computer Engineering at STES's Smt. Kashibai Navale College of Engineering, Pune, India.

Dr. Patrick Ohemeng Gyaase holds a Ph.D. (Information Technology) from Aalborg University, Denmark: Center for Communication, Media and Information Technologies, Department of Electronics. He is currently a Senior Lecturer at the Faculty of Information, Communication Sciences and Technology (ICST), Catholic University College of Ghana, Fiapre, Sunyani.

Dr. Pollyanna Rodrigues Gondin, Obtained a M.A. in Economics from Federal University of Uberlândia. Ph.D. and Postdoctoral Researcher in the Graduate Program in Public Policies at Federal University of Paraná, Brazil. Member of the Brazilian Association of Industrial Economics and Innovation. Research areas: industrial organization and industrial studies, with emphasis on the following topics: micro and small enterprises, clusters, Systems of Innovation (SI), IOT and industrial policy.

Dr. Prashant S. Dhotre has obtained his B.E. degree in Computer Science and Engineering from Shri Ramanand Teertha Marathwada University, Nanded, India in 2004 and M.E. degree in Information Technology from Savitribai Phule Pune University, Pune, India. In 2017 awarded PhD in Wireless Communication, Department of Electronic Systems, Center for Communication, Media and Information Technologies Copenhagen, Aalborg University Denmark. He has more than 13 years of teaching and research experience. He has authored 2 books on a subject "Theory of Computations" for UG students of Computer Engineering. He loves to interact with the students through a series of guest lectures on "Design and Analysis of Algorithms", "Theory of Computation", and "Data Structures" in various Engineering Colleges across the Pune region. Currently, he is working as Assistant Professor in Department of Computer Engineering, STES's Sinhgad Institute of Technology and Science, Pune, India.

Dr. Roslyn Layton promotes evidence-based policy for the international ICT sectors. She is one of the few academics who has performed empirical research on policies such as net neutrality and zero rating. She served on the President Elect Transition Team for the Federal Communications Commission (FCC) and has testified to the United States Senate Committee on the Judiciary on General Data Protection Regulation (GDPR). She is a Visiting

Researcher at the Center for Communication, Media, and Information Technologies (CMI) at Aalborg University in Copenhagen, Denmark and a Visiting Scholar at the American Enterprise Institute. Roslyn holds a PhD from the Doctoral School of Engineering and Science at Aalborg University in Denmark, an MBA from the Rotterdam School of Management (where she won the *Wall Street Journal Europe* Women in Business Scholarship), and BA in International Service from American University. Roslyn has worked in the IT, financial, and biotech sectors in the US, EU, and India. She writes for Forbes, US News and World Report, and AEIdeas.

Samuel Agbesi is a Ph.D. fellow at Center for Communication, Media and Information Technology (CMI), Aalborg University, Copenhagen. He has over fifteen years' experience in Information and Communication Technology, with specialization in Database Administration and Software Engineering. His area of research is in Usability engineering, and Blockchain & privacy.

Dr. Silvia Elaluf-Calderwood is a professional consultant and researcher with academic and industrial experience in mobile technology research, business models analysis, telecommunications strategy, mobile payments, big data in the UK, the Netherlands and in multidisciplinary projects at EU/International level (USA, Latin America and Asia) in socio technical areas. Industry qualified engineer (CCNP, CCNA, MSc(Eng), BSc) with a telecoms and Internet background (networking, economic models and IP). She currently holds dual positions as a Visiting research Scholar at Florida International University (FIU, USA) and Adjunct Professor at the University of Syracuse (NY, USA). In the past she has been an Associated research fellow at Oxford Brookes University (UK) and Research fellow at the London School of Economics and Political Science (LSE, UK). She obtained her PhD at the Information Systems at the Department of Management at the LSE. Specialization Mobile technology, distributed working practices and organizational change. For 2018–2019 she is the holder of the Catedra Telefonica IBEI Research Fellowship (Barcelona, Spain). She has published a number of book chapters and actively engages in activities related to the Internet Governance Forum (IGF), Internet Society (UK chapter), BCS (British Computer Society) and other social blogs and media (both academic and business related such as the Brookings Institute). Currently she is teaching Distributed management Systems (cloud computing) at FIU and Information Policy at Syracuse.

Dr. Stephane Nwolley, Jnr has 13+ years of experience in the ICT industry, as an academic and entrepreneur, Stephane has worked with MTN and Huawei as an OSS engineer. His experience and hands on knowledge of the industry helps in the implementation of efficient solutions that aid in delivering the best of services to customers. Formerly, the CEO of Nalo Solutions Limited, he was the product lead for project with clients including Ecobank, GIPC, UBA, amongst others. He holds a PhD in ICT Management (Big Data). He is skilled in synthesizing large amounts of information into insights for strategy, innovation and business development. He is also an adjunct lecturer in the computer science department of Ashesi University.

Dr. Walter Shima holds a Ph.D. from Institute of Economics at Federal University of Rio de Janeiro. Professor in the Graduate Program in Public Policies at Federal University of Paraná, Brazil. Member of the Brazilian Association of Industrial Economics and Innovation. Research areas: industrial organization, IOT, industrial policy and enterprise-university relationship.

About the Editors

Prof. Knud Erik Skouby is professor and founding director of center for Communication, Media and Information technologies, Aalborg University, Copenhagen – a center providing a focal point for multi-disciplinary research and training in applications of CMI. Has a career as a university teacher and within consultancy since 1972. Working areas: *Techno-economic Analyses; Development of mobile/wireless applications and services: Regulation of telecommunications.*

Project manager and partner in a number of international, European and Danish research projects. Served on a number of public committees within telecom, IT and broadcasting; as a member of boards of professional societies; as a member of organizing boards, evaluation committees and as invited speaker on international conferences; published a number of Danish and international articles, books and conference proceedings. Editor in chief of Nordic and Baltic Journal of Information and Communication Technologies (NBICT); Board member of the Danish Media Committee. Chair of WGA in Wireless World Research Forum; Dep. chair IEEE Denmark. Member of the Academic Council of the Technical Faculty of IT and Design, AAU.

Dr. Idongesit Williams is a Post-Doctoral Research fellow at Aalborg University. He holds a Ph.D in Information and Communication Technologies, specializing in telecommunication and Internet infrastructure development and policies. He has for 7 years worked as a Lecturer at Aalborg University Copenhagen where he taught Internet Governance and Economics. Currently, he is involved in EU projects. His current project is on the promotion of digitization in the Baltic Sea region. In addition to his core area of specialization, he has conducted research into other areas involving the diffusion and adoption of ICTs in organizations and by Individuals. He has published widely with more than 40 publications on research areas ranging from Public Private Partnerships in the development of Broadband infrastructure, ICT business modeling, Knowledge management and organizational learning, Gender, ICT, digitalization in the EU and entrepreneurship among others. He has served as Special issue guest editor to journals. He is also a consultant for SMEs on the development of requirement specifications and use cases for IT based services.

Dr. Albert Gyamfi obtained his PhD from Aalborg University, Copenhagen, Denmark. Where he conducted a study to investigated the usage of Web 2.0 applications on Knowledge Management Processes, using the Ghanaian Cocoa Sector as a case. He earned his Masters degree in Management Information System from the University of Ghana. He's currently pursuing Masters in Computer Science Degree, at the University of Regina in Canada, with a research focus on Knowledge-based System approach to Internet of Things (Smart Cities, Smart Factories and Smart Agriculture) with keen interest in 5G technologies. He also has a general interest in the theoretical foundations of Information Systems, with specific focus on knowledge management systems and knowledge-based cognitive systems.